HEALTH IMPROVEMENT AND POVERTY REDUCTION MANAGEMENT

Sonsoles Rodríguez González
Family and Community Doctor

Laureano Rodríguez Navarro
BA in Tourism

ISBN 978-1-291-02783-9
© 2012 All rights reserved

July 2012

Foreword

One in five people on the planet – two-thirds of them women – live in abject poverty. While the last century saw great progress in reducing poverty and improving well-being, poverty remains a global problem of huge proportions. Of the world's 6 billion people, 2.8 billion live on less than $2 a day, and 1.2 billion on less than $1 a day. To address this challenge, the world's governments committed themselves at the United Nations Millennium Summit to the Millennium Development Goals, including the overarching goal of halving extreme poverty by the year 2015.

Yet, our planet's capacity to sustain us is eroding. The problems are well-known – degrading agricultural lands, shrinking forests, diminishing supplies of clean water, dwindling fisheries, and the threat of growing social and ecological vulnerability from climate change and loss of biological diversity. While these threats are global, their impacts are most severe in the developing world – especially among people living in poverty who have the least means to cope.

Is this environmental decline inevitable in order for poverty to be reduced? We argue not. Indeed, quite the opposite is true. If we do not successfully arrest and reverse these problems, the world will not be able to meet the Millennium Development Goals, particularly the goal of halving extreme poverty. As this paper demonstrates, tackling environmental degradation is an integral part of effective and lasting poverty reduction. The 2002 World Summit on Sustainable Development (WSSD) provides the international community with a pivotal opportunity to redirect the global debate, and to forge a more integrated and effective global response to poverty and environmental decline.

To succeed, we need to focus on the most important links between poverty, the environment and sustainable development. For many, ensuring sound environmental management means curtailment of economic opportunities and growth, rather than their expansion … too often; it is viewed as a cost rather than an investment. There is no simple relationship between economic growth and environmental degradation, but appropriate policies nationally and internationally can have a major impact on both fronts. To this end, we need to look beyond what environmental institutions can do, and search for opportunities across all sectors.

Contents

OVERVIEW: POVERTY REDUCTION, HEALTH AND ENVIRONMENTAL MANAGEMENT

I

PART 1: WHY THE ENVIRONMENT MATTERS TO PEOPLE LIVING IN POVERTY

1.1 Livelihoods and the environment
1.2 Health and the environment
1.3 Vulnerability and the environment
1.4 Economic growth and the environment

PART 2: POLICY OPPORTUNITIES TO REDUCE POVERTY AND IMPROVE THE ENVIRONMENT

2.1 Improving governance
2.2 Enhancing the assets of the poor
2.3 Improving the quality of growth
2.4 Reforming international and industrialized country policies
CONCLUSION
ENDNOTES
ABBREVIATIONS AND ACRONYMS
REFERENCES

OVERVIEW
Linking Poverty Reduction
And Environmental Management

Addressing environmental issues that matter to the poor is critical to sustained poverty reduction and achieving the Millennium Development Goals ... But this requires a more 'pro-poor' and integrated approach – linking action at local, national and global levels.

Prepared as a contribution to the 2002 World Summit on Sustainable Development (WSSD), POVERTY REDUCTION, HEALTH AND ENVIRONMENTAL MANAGEMENT focuses on ways to reduce poverty and sustain growth by improving management of the environment, broadly defined. It seeks to draw out the links between poverty and the environment, and to demonstrate that sound and equitable environmental management is integral to achieving the Millennium Development Goals, in particular eradicating extreme poverty and hunger, reducing child mortality, combating major diseases, and ensuring environmental sustainability.

Four priority areas for sustained policy and institutional change are highlighted:

☐ Improving governance for pro-poor and pro-environment policies, institutions and services, with particular attention to the needs of women and children;

☐ Enhancing the assets of the poor and reducing their vulnerability to environment-related shocks and conflict;

☐ Improving the quality of growth to protect the asset base of the poor and expand opportunities for sustainable livelihoods;

☐ Reforming international and industrialized country policies related to trade, foreign direct investment, aid and debt.

Policy opportunities exist to reduce poverty and improve the environment

The environment matters greatly to people living in poverty. The poor often depend directly on natural resources and ecological services for their livelihoods; they are often the most affected by unclean water, indoor air pollution and exposure to toxic chemicals; and they are particularly vulnerable to environmental hazards such as floods and prolonged drought, and to environment-related conflict. Addressing these poverty-environment linkages must be at the core of national efforts to eradicate poverty.

Many policy opportunities exist to reduce poverty by improving the environment – but there are significant and often deeply entrenched policy and institutional barriers to their widespread adoption. The past decade of experience since the 1992 Earth Summit in Rio reveals some important lessons that help point the way forward. Three broad lessons are highlighted here:

☐ First and foremost, poor people must be seen as part of the solution – rather than part of the problem. Efforts to improve environmental management in ways that contribute to sustainable growth and poverty reduction must begin with the poor themselves. Given the right incentives and support – including access to information and participation in decision-making – the poor will invest in environmental improvements to enhance their livelihoods and well-being. At the same time, however, it is essential to address the activities of the non-poor since they are the source of most environmental damage.

☐ The environmental quality of growth matters to the poor. Environmental improvement is not a luxury preoccupation that can wait until growth has alleviated income poverty, nor can it be assumed that growth itself will take care of environmental problems over the longer-term as a natural by-product of increasing affluence. First, this ignores the fundamental importance of environmental goods and services to the livelihoods and well-being of the rural and urban poor. Second, there are many examples of how bad environmental management is bad for growth, and of how the poor bear a disproportionate share of the costs of environmental degradation. Ignoring the environmental soundness of growth – even if this leads to short-run economic gains – can undermine long-run growth and its effectiveness in reducing poverty.

☐ Environmental management cannot be treated separately from other development concerns, but requires integration into poverty reduction and sustainable development efforts in order to achieve significant and lasting results. Improving environmental management in ways that benefit the poor requires policy and institutional changes that cut across sectors and lie mostly outside the control of environmental institutions – changes in governance, domestic economic policy, and in international policies.

Improving governance

☐ Integrate poverty-environment issues into nationally-owned poverty reduction strategies, including macroeconomic and sect oral policy reforms and action programmes, so that they can become national sustainable development strategies.

☐ Engage poor and marginalized groups in policy and planning processes to ensure that the key environmental issues that affect them are adequately addressed, to build ownership, and to enhance the prospects for achieving lasting results.

☐ Address the poverty-environment concerns of poor women and children and ensure that they are given higher priority and fully integrated into poverty reduction strategies and policy reforms – for example, the growing burden of collecting scarce water and fuelwood supplies, and the effects of long-term exposure to polluted indoor air.

☐ Implement anti-corruption measures to counter the role of corruption in the misuse of natural resources and weak enforcement of environmental regulations – for example, the destructive impacts of illegal logging and unregulated mining, or the preference for construction of new power and water investments over increasing the efficiency of existing investments.

☐ Improve poverty-environment indicators to document environmental change and how it affects poor people, and integrate into national poverty monitoring systems. This should be complemented by measures to improve citizens' access to environmental information.

Enhancing the assets of the poor

☐ Strengthen resource rights of the poor by reforming the wider range of policies and institutions that influence resource access, control and benefit-sharing, with particular attention to resource rights for women. This includes central and sub-national government, traditional authorities, the legal system, and local land boards, commissions and tribunals.

☐ Support decentralization and local environmental management – land, water and forest resource management, and provision of water supply and sanitation services – by strengthening local management capacity and supporting women's key roles in managing natural resources.

☐ Expand access to environmentally-sound and pro-poor technology, such as crop production technologies that conserve soil and water and minimize the use of pesticides, or appropriate renewable energy and energy efficient technologies that also minimize air pollution. This includes support for indigenous technologies, and the need to address the social, cultural, financial and marketing aspects of technical change.

☐ Promote measures that reduce the environmental vulnerability of the poor by strengthening participatory disaster preparedness and prevention capacity, supporting the formal and informal coping strategies of vulnerable groups, and expanding access to insurance and other risk management mechanisms.

☐ Reduce the vulnerability of the poor to environment-related conflict by improving conflict resolution mechanisms in the management of natural resources and addressing the underlying political issues that affect resource access.

Improving the quality of growth

☐ Integrate poverty-environment issues in economic policy and decision-making by strengthening the use of environmental assessment and poverty social impact analysis.

☐ Improve environmental valuation at both the macro and micro level, in order to highlight the full cost of environmental degradation for the poor in particular and the economy in general, and to improve economic decision-making.

☐ Expand private sector involvement in pro-poor environmental management to maximize the efficiency gains from private sector participation, while safeguarding the interests of the poor. This requires capacity within government to negotiate with the private sector – for example, to ensure that utility privatization benefits the poor – and to forge effective public-private partnerships that enhance the poor's access to environmental services.

☐ Implement pro-poor environmental fiscal reform including reform of environmentally-damaging subsidies, improved use of rent taxes to better capture and more effectively allocate resource revenues, and improved use of pollution charges to better reflect environmental costs in market prices.

Reforming international and industrialized country policies

☐ Reform trade and industrialized country subsidy policies to open up markets to developing country imports while avoiding environmental protectionism, and to reduce subsidies that lead to unsustainable exploitation – such as subsidies for large-scale commercial fishing fleets that encourage over-harvesting in developing country fisheries.

☐ Make foreign direct investment more pro-poor and pro-environment by encouraging multinational corporations to comply with the revised OECD Code of Conduct for Multinational Enterprises, and to report on the environmental impact of their activities in line with the UN Environment Programme's Global Reporting Initiative.

☐ Increase funding for the Global Environment Facility as the major source of funding for global public goods in the environment, such as a stable climate, maintenance of biodiversity, clean international waters and the protective ozone layer. These benefit the whole world as well as the poor themselves – so the rich world must pay a fair share for their maintenance.

☐ Enhance the contribution of multilateral environmental agreements (MEAs) to national development objectives by strengthening developing country capacity to participate in the negotiation and implementation of MEAs (for example, to ensure that the Clean Development Mechanism promotes investments that benefit the poor). Also, improved coordination is needed between MEAs so that scarce developing country capacity is used most effectively.

☐ Encourage sustainable consumption and production – industrialized country consumers and producers through their trade, investment, pollution emissions and other activities affect the environmental conditions of developing countries. Making rich country consumption and production more sustainable will require a complex mix of institutional changes – addressing market and government failures as well as broad public attitudes.

☐ Enhance the effectiveness of development cooperation and debt relief with more priority for poverty-environment issues, particularly for the poorest countries where aid and debt relief continue to have a valuable role to play in helping governments to make many of the changes recommended above. Mainstream environment in donor agency operations through staff training, development and application of new skills, tools and approaches, and revisions to the way resources and budgets are allocated. Transparent monitoring of progress against stated objectives and targets is needed in order to hold development agencies accountable and to ensure that a commitment by senior management to addressing poverty-environment issues is put into practice throughout the organization.

Conclusion

This paper looks ahead with some degree of hope and optimism for the future – there are sometimes win-win opportunities, and there are rational ways of dealing with trade-offs. Environmental degradation is not inevitable, nor the unavoidable result of economic growth. On the contrary, sound and equitable environmental management is key to sustained poverty reduction and achievement of the Millennium Development Goals. There are significant policy opportunities to reduce poverty and improve the environment, but more integrated and pro-poor approaches are needed. The World Summit on Sustainable Development is an opportunity to focus on what is most important and to forge a coherent framework for action, with clear goals and achievable targets backed-up by adequate resources and effective and transparent monitoring mechanisms. There can be no more important goal than to reduce and ultimately eradicate poverty on our planet.

PART 1
Why the Environment Matters
to People Living in Poverty

"Water is life and because we have no water, life is miserable" (Kenya)
"We think the earth is generous; but what is the incentive to produce more than the family needs if there are no access roads to get produce to a market?" (Guatemala)
"In the monsoons there is no difference between the land in front of our house and the public drain. You can see for yourself" (India)

In their own words, the environment matters greatly to people living in poverty. Indeed, poor people's perceptions of well-being are strongly related to the environment in terms of their livelihoods, health, vulnerability, and sense of empowerment and ability to control their lives. Figure 1 provides a simplified framework for understanding how environmental management relates to poverty reduction, and why these poverty-environment linkages must be at the core of action to achieve the Millennium Development Goals and related national poverty eradication and sustainable development objectives.

FIGURE 1: ENVIRONMENT AND THE MILLENNIUM DEVELOPMENT GOALS
Environmental management for poverty reduction   Dimensions of poverty
        Development goals

Part 1 of the paper focuses on the poverty-environment relationship by examining how environmental conditions in both rural and urban settings relate to three key dimensions of human poverty and well-being:

- Livelihoods – poor people tend to be most dependent upon the environment and the direct use of natural resources, and therefore are the most severely affected when the environment is degraded or their access to natural resources is limited or denied;

- Health – poor people suffer most when water, land and the air are polluted;

- Vulnerability – the poor are most often exposed to environmental hazards and environment-related conflict, and are least capable of coping when they occur.

We also are concerned with the relationship between growth and the environment and how it affects the poor and efforts to reduce poverty. The environmental soundness of growth matters considerably to the poor, and countries with similar levels of income and growth can have quite different levels of environmental performance.

While Figure 1 illustrates the main pathways between environmental conditions and dimensions of poverty, in reality these linkages are multi-dimensional, dynamic and often inter-connected:

- Poverty is now widely viewed as encompassing both income and non-income dimensions of deprivation – including lack of income and other material means; lack of access to basic social services such as education, health and safe water; lack of personal security; and lack of empowerment to participate in the political process and in decisions that influence one's life. The dynamics of poverty also are better understood, and extreme vulnerability to external shocks is now seen as one of its major features. Environment refers to the biotic and abiotic components of the natural world that together support life on earth – as a provider of goods (natural resources) and ecosystem services utilized for food production, energy and as raw material; a recipient and partial recycler of waste products from the economy; and an important source of recreation, beauty, spiritual values and other amenities.

- The nature and dynamics of poverty-environment linkages are context-specific – reflecting both geographic location and economic, social and cultural characteristics of individuals, households and social groups. Different social groups can prioritize different environmental issues (Brocklesby and Hinshelwood, 2001). In rural areas, poor people are particularly concerned with their access to and the quality of natural resources, especially water, crop and grazing land, forest products and biomass for fuel. For the urban poor, water, energy, sanitation and waste removal are key concerns. Poor women regard safe and physically close access to potable water, sanitation facilities and abundant energy supplies as crucial aspects of well-being, reflecting their primary role in managing the household.

☐   Environmental management, as used in this paper, extends well beyond the activities of public environmental institutions. In relation to poverty, environmental management is concerned fundamentally with sustaining the long-term capacity of the environment to provide the goods and services upon which people and economies depend. This means improving environmental conditions and ensuring equitable access to environmental assets – in particular land and biological resources, and safe and affordable water supply and sanitation – in order to expand poor people's livelihood opportunities, protect their health and capacity to work, and reduce their vulnerability to environment-related risks. This broader conception of poverty and environment, and of environmental management, is essential to understanding the linkages between them and to identifying appropriate policy and institutional options for improving these linkages.

There have been some impressive gains since the 1972 United Nations Conference on the Human Environment – the first global conference devoted to environment and development issues. There has been a proliferation of environmental policies and institutions at national and sub-national levels, and environmental issues are firmly placed on the agendas of governments, civil society and the private sector. Major global environmental agreements have been forged and global environmental organizations established. Environmental sustainability has become a core concern of bilateral and multilateral development cooperation, and billions of dollars have been spent on environment-related programmes and projects.

Tangible progress also has been achieved 'on the ground', although the picture is usually mixed. For example, in the 1990s some 900 million people gained access to improved water sources. However, this was merely enough to keep pace with population growth, and about 1.2 billion people are still without access to improved water sources, with rural populations particularly under-served (Devarajan et al, 2002). Another example is the productivity of soil used for cereal production, which increased on average in developing countries from 1979-81 to 1998-2000. However, it fell in some 25 countries, most of them in

Africa, with land degradation being one factor behind the decline (World Bank, 2002c).

Despite these gains, pressure on the environment continues to mount worldwide, posing major challenges to the prospects for poverty reduction and human development in developing countries, in particular the least developed countries. The situation is summed-up succinctly in UNEP's 2002 Global Environment Outlook report: "…The level of awareness and action has not been commensurate with the state of the global environment today; it continues to deteriorate" (UNEP, 2002). Box 1 summarizes key environmental challenges facing developing countries in relation to the Millennium Development Goals. These linkages are addressed in more detail in the following sections on livelihoods, health, vulnerability and growth.

BOX 1: KEY ENVIRONMENTAL CHALLENGES TO THE MILLENNIUM DEVELOPMENT GOALS

Goal 1: Eradicate extreme poverty and hunger
☐ The livelihood strategies and food security of the poor often depend directly on healthy ecosystems and the diversity of goods and ecological services they provide. Land degradation, deforestation and biodiversity loss undermine ecosystem health and productivity, and in many areas threaten the poor's access to environmental resources.
☐ The poor are particularly vulnerable to environmental shocks and stresses. Environmental degradation is exacerbating the frequency and impact of droughts, floods, forest fires and other natural hazards. Projected impacts from global climate change, such as shifting agricultural zones and rising sea levels, are predicted to be most severe in developing countries and will disproportionately affect poor communities that have limited capacity to cope or adapt, in particular agricultural and coastal populations.

Goal 2: Achieve universal primary education
☐ Children can spend significant time helping with household tasks such as water and fuelwood collection when environmental resources are scarce, reducing time spent at school.
☐ When environmental pressures undermine household livelihood security, children often must leave school entirely in order to help the family to cope.

Goals 4 & 6: Reduce child mortality; Combat HIV/AIDS, malaria and other major diseases
☐ Lack of access to safe drinking water and sanitation, and exposure to indoor and outdoor air pollution and toxic chemicals, are major causes of ill-

health and life-threatening disease in developing countries, particularly among poor women and children.

☐ More than 1 billion people do not have access to safe drinking water, and more than 2 billion people lack access to adequate sanitation and clean energy services.

Source: UNEP (2002).

1.1 Livelihoods and the environment

"There is a strong correlation between sound natural resource management and poverty reduction." (Cambodia Interim Poverty Reduction Strategy Paper, 2000)

"Environmental protection is of significant relevance to poverty reduction not only because . . . the poor are disproportionately exposed to the impact of deteriorating environmental conditions, but also because the low productivity of resources is determined by the poverty of rural people. Progress in this area depends on public recognition of the mutual links between environmental issues, economic development and the level of poverty." (Armenia Interim Poverty Reduction Strategy Paper, 2001)

The poor, particularly those living in rural areas, often rely on natural resources and ecosystem services) for their livelihoods. Increasingly, the rural poor live in areas of high ecological vulnerability and relatively low levels of biological or resource productivity, such as subtropical drylands or steep mountain slopes. Estimates indicate that there may be almost twice as many rural poor living on such 'marginal' lands as on favored lands in developing countries (CGIAR et al., 1997). If current trends in poverty and environmental degradation persist, by 2020 more than 800 million people could be living on marginal lands (Hazell and Garrett, 1996). Limited access to land and other natural resources is another key aspect of rural poverty – more than half of the rural poor have landholdings too small to provide an adequate income, and nearly a quarter are landless (UNCHS, 1996). Thus, both environmental conditions and access to natural resources are crucial to the ability of poor people to sustain their livelihoods.

Natural resources

Natural resources can be a primary source of livelihood or may supplement the household's daily needs and income. A growing body of research shows that poor rural households often derive a significant share of their incomes from natural resources (see Box 2). A study of 80 villages in dry zone India found that common property resources provided 15-23 percent of total income for the poorer households of Andhra Pradesh, but only 1-3 percent for the larger farm households (Jodha, 1986).

Natural resource degradation is undermining the livelihoods and future livelihood opportunities of large numbers of the poor. This is most evident with respect to agricultural systems. Soil and water degradation is a major threat to improving agricultural productivity, which underpins the livelihoods of the vast majority of the rural poor and is a cornerstone of poverty reduction strategies in many countries. International trade can cushion local deficiencies in food production for the better off, but for the poorest this is not an option.

Current estimates are that up to one billion people are affected by soil erosion and land degradation due to deforestation, over-grazing and agriculture. Water scarcity is a major issue in more than 20 developing countries. If current trends in water use persist, two-thirds of the world's population could be living in countries experiencing moderate or severe water scarcity by 2025. Fisheries provide livelihoods for some of the poorest and most marginalized groups, and often are the main source of animal protein for the poor. Yet, many small-scale fisheries are over-harvested, often by commercial enterprises that do not benefit the poor (IFAD, 2001; WRI, 2000, UNEP, 2002).

Poor people are affected by natural resource degradation much more than the better off because of their limited assets and their greater dependence on common property resources for their livelihoods. For example, better-off farmers are able to compensate for falling natural fertility by using more fertilizer, but fertilizer use by poor people is very low. Under these circumstances, land degradation has been shown to have pernicious direct effects upon poverty. In a study in West Africa, children showing growth abnormalities associated with poor nutrition (stunting) were found most frequently in areas of high soil degradation (GRID/Arendal, 1997).

BOX 2: NATURAL RESOURCES AND HOUSEHOLD INCOME IN RURAL AREAS OF ZIMBABWE

A study of 29 villages in southern Zimbabwe shows that natural resources account for about one-third of average total household income (see below). The study also revealed that the poorer the household, the greater the share of income from natural resources and the greater the dependence on common property resources.

Over 2 billion people continue to rely on biomass fuels and traditional technologies for cooking and heating, and 1.5-2 billion people have no access to electricity (UNDP, UNDESA and World Energy Council, 2000). Shortage of wood fuel imposes time and financial costs on poor households, putting a particular burden on households that are short of labour and making it harder for children to attend school.

Poor rural women are disproportionately affected by natural resource degradation because of their particular dependence on common property resources. Participatory poverty assessments and other studies have shown the increased time, physical burden and personal risk that women face in having to travel greater distances in order to collect fuel, fodder and water due to growing resource scarcity or more restricted access to common property areas. This reduces the time spent on income-generating activities, crop production, and household and child-rearing responsibilities (Brocklesby and Hinshelwood, 2001; Dasgupta and Das, 1998).

Ecosystem services

Ecosystems – such as forests, agroecosystems, grasslands, freshwater and coastal ecosystems (including coral reefs) – provide essential 'services' that contribute in numerous ways to productive activities. These services are 'public goods', providing indirect values that are only partially traded in the market place, but which are vital to the livelihoods of the poor, especially in more marginal environments or where the poor have limited access to external technology and other inputs (Koziell and Saunders, 2001). By maintaining productivity and a healthy and stable environment, ecosystem services also contribute to maintaining livelihood options and the potential for livelihood diversification. When ecosystem functions are impaired, this inevitably leads to a narrowing of livelihood choices and an increase in the vulnerability of the poor.

While ecosystems can be highly resilient to human disturbances, certain ecosystem types are at particular risk of a sudden collapse. In particular, coral reefs, freshwater systems and nutrient-poor lands may go from a functioning to a non-functioning state in a very short time due to pollution, overuse or other perturbations that exceed a certain threshold. The consequence is that people who are dependant on these ecosystems may find themselves deprived of essential goods and services in a relatively short time span and unable to cope or adapt.

Some examples of ecosystem services that support livelihoods include: provision of natural habitat for wild pollinators that are essential to food crops, natural predators that control crop pests and soil organisms important to agricultural productivity; watershed protection and hydrological stability, including recharging of water tables and buffering of extreme hydrological conditions which might otherwise precipitate drought or flood conditions; maintenance of soil fertility through storage and cycling of essential nutrients; and breakdown of waste and pollutants.

1.2 Health and the environment

"…a study in Tegucigalpa showed…high lead intoxication in the children attending public schools. The study also notes that contaminants in soil and water are responsible for a high index of diarrhea diseases… Soil and water pollution is further compounded by solid waste dumping with low coverage of garbage collection services, poor waste management, and the lack of sanitary landfills. Respiratory diseases are also common, especially among children under five…partly caused by increasing number of cars and the presence of factories that are not subject to any kind of environmental regulations." (Honduras Poverty Reduction Strategy, 2001)

Up to one-fifth of the total burden of disease in the developing world – and up to 30% in sub-Saharan Africa – may be associated with environmental risk factors. This is comparable to malnutrition and larger than any other preventable risk factors and groups of disease causes. While the total burden of disease in poor countries is about twice that of rich countries, the disease burden from environmental risks is 10 times larger in poor countries (see Figure 2). The poor, particularly women and children, are most affected by environmental health problems, and traditional environmental hazards – lack of safe water and sanitation, indoor air pollution and exposure to disease vectors – play by far the largest role (Lvovsky, 2001; WHO, 1997). Indeed, poor people are acutely aware of how poor environmental health affects their ability to move out of poverty (Brocklesby and Hinshelwood, 2001; Narayan, 2000).

Analyzing the impact of policy changes and investments on the poor is important in bringing out the specifics in the trade-off between income growth and environmental quality. Such analysis frequently shows that the poor stand to benefit from environmental interventions now rather than later. Many interventions are low-cost, yet can save people from disease that can seriously impair their earning capability and welfare.

FIGURE 2: BURDEN OF DISEASE AND ENVIRONMENTAL RISKS (1990)

Note: Disability-Adjusted Life Years (DALYs) are a measure of the burden of disease. They reflect the total amount of healthy life lost to all causes, whether from premature mortality or from some degree of disability during a period of time.

Source: Lvovsky (2001).

Water and sanitation

Inadequate access to safe drinking water and sanitation, combined with poor hygiene practices, are major causes of ill-health and life-threatening disease in developing countries. The rural poor rely on natural water sources such as streams for their washing and drinking water (see Box 3). Water-related diseases, such as diarrhea and cholera, kill an estimated 3 million people a year in developing countries, the majority of which are children under the age of five (Murray and Lopez, 1996).

Vector borne diseases such as malaria account for up to 2.5 million deaths a year, and are linked to a range of environmental conditions and factors related to water contamination and inadequate sanitation (WRI, 1998). These are likely to worsen as a result of climate change (IPCC, 2001).

BOX 3: BURDEN OF WATER COLLECTION ON WOMEN AND CHILDREN

A recent water use study in Kenya, Uganda and Tanzania went back to the same 34 sites that were studied in 1972. Water is still primarily collected by women and children and carried on the head leading to headaches, general fatigue and pains in the chest, neck and waist. The distance walked to collect water was about 580 m in rural areas (although for some it can reached over 4 km) and 300 m in urban areas. This is a slight improvement since 1972 due to more standpipes, wells and private vendors, including in rural areas. However, due to population increase, time spent queuing has increased significantly, especially in urban areas. A return journey to collect water takes about 25 minutes (double the time since 1972), and 3.9 trips per day are made by each

household. Thus, an average household spends 1 hour and 40 minutes collecting water each day. This reduces time for cooking and can reduce the amount of time children spend at school.
Source: IIED et al. (2002).

Pollutants

Indoor air pollution caused by the burning of traditional biomass fuels (wood, dung, crop residues) for cooking and heating affects one billion people, killing an estimated 2 million women and children each year (Smith, 1999). In India, recent studies suggest that 130,000–150,000 women may die prematurely as a result of indoor air pollution (Smith, 2000). A recent study of rural households in central Kenya found that "exposure to high emissions from cooking and other domestic activities for adults result in women being twice as likely as men to be diagnosed with acute respiratory infection or acute lower respiratory infections" (Ezzati and Kammen, 2001). This has been confirmed by similar studies in Gambia (Campbell, 1997) and Guatemala (Bruce et al., 1998). In addition, the increased time and energy involved in the collection of biomass fuels contributes to the physical burden and ill-health of women and children.

Outdoor air pollution is becoming a more significant health issue in urban areas of a number of developing countries, especially in large middle income countries such as China and India, and is projected to become as important a health risk as indoor air pollution over the next two decades.

Pesticide poisoning is a significant health problem among poor farmers in developing countries, although the exact extent is not well documented. One estimate by WHO in 1990 indicated a level of some 3 million cases of acute, severe poisoning per year worldwide. Widening the scope to cases of pesticide "exposure" that can either result in acute illness or chronic health impacts, estimates for Africa alone point to some 11 million cases per year (Goldman and Tran, 2002). The poor also suffer more indirect effects from excessive use of pesticides, such as depletion of fish stocks due to pesticide loads in agricultural runoff.

1.3 Vulnerability and the environment

"Natural disasters are a risk factor, which affect the pace of economic growth and destroy the assets of the poorest segments of the population in affected areas, reducing them to a state of dependency, at least temporarily, on donations ... natural disasters seriously affect the living conditions of affected populations, and constitute an obstacle to a definite break with certain degrees and patterns of poverty. Therefore, measures aimed at managing this risk are of the utmost importance." (Mozambique Action Plan for the Reduction of Absolute Poverty, 2001-2005)

Insecurity is one of the key concerns of poor people, including their vulnerability to unpredictable events. Insecurity relates to people's risk of exposure, susceptibility to loss, and capacity to recover. Both the rural and urban poor are most often exposed to environmental hazards and environment-related conflict, they suffer the greatest losses (at least in relative terms) and they are in the weakest position to cope and adapt.

Environmental stresses and shocks

Resource mismanagement and environmental degradation can exacerbate the frequency and impact of droughts, floods, forest fires and other natural hazards. The poor are the most vulnerable to environmental disasters ('shocks') as well as more gradual processes of environmental degradation ('stresses') – as the majority of the rural poor live in ecologically-fragile areas, while the urban poor often live and work in environments with a high exposure to environmental hazards. By exacerbating economic deprivation in the short term, environmental disasters can compromise long-term welfare by forcing affected households to sell-off assets that would otherwise be used to meet future needs and contingencies. The effects of droughts and long-term land degradation are more gradually felt. They may build up over several years, during which a household's accumulated reserves are run down as a result of recurrent years of poor production. This will result in a slow but inexorable inability to invest in production and often leads to impoverishment and the abandonment of land.

Natural hazards claim an estimated 100,000 lives each year, and inflict billions of dollars in damage. While natural hazards can strike everywhere, about 97 percent of the natural-disaster related deaths occur in developing countries. The relative economic losses are also highest in poor countries. (ISDR, 2002). Natural disasters affected an estimated 256 million people in developing countries in 2000 (ICRC, 2001).

When asked, the poor talk of living in increasingly fragile biophysical contexts (in both urban and rural settings) and experiencing natural hazards, changing climatic conditions and unpredictable seasons. These environmental stresses were making livelihood tasks more time-consuming, more dangerous, more costly and often requiring more inputs. Poor people highlight their dependence on common property resources – grazing lands, water bodies and forests –

often as a safety net. A decline in these resources increases their vulnerability (Brocklesby and Hinshelwood, 2001).

Increasingly, environmental degradation and disasters cause their victims to migrate in search of better conditions. People may be able to recover, with help, from sudden disasters, and people often return and re-build after floods and storms. However, long-term attrition caused by drought or land degradation has led to permanent migration from susceptible areas such as the Sahel. The Red Cross estimates that 1998 was the first year in which the number of refugees from environmental disasters exceeded those displaced as a result of war (ICRC, 1999). However, much of the information on environmental degradation and disasters as a source of migration is anecdotal, and it is difficult to analyze the complex system of inter-connected social, demographic and environmental phenomena that together form the basis for cross-border migration (Leighton, 1999).

The frequency, intensity and duration of extreme weather events is likely to increase as a result of climate change. The latest report on the impacts of climate change suggest that many developing countries in Africa, Asia and Latin America will suffer potentially significant negative impacts from increased food insecurity, greater spread of vector borne disease, more flooding and exacerbation of land degradation (see Box 4).

Poor people employ a range of coping mechanisms and survival strategies in the face of environmental degradation and disasters. But their capacity to mitigate and recover from disaster is often constrained by the wider institutional context, in addition to factors related to their social and economic status. For example, in many developing countries, there is a lack of social safety nets and other protections that can help to mitigate the impacts of environmental disasters on the poor. Informal institutions such as local social networks also are important, and their density and capacity can underpin the ability of the poor to cope.

BOX 4: IMPACTS OF GLOBAL CLIMATE CHANGE ON THE POOR

Climate change impacts will particularly affect poor countries who will find adaptation measures more costly, and affect poor people who have more limited coping mechanisms. Major impacts include declining water availability, reduced agricultural productivity, spread of vector borne diseases to new areas, increased flooding from sea level rise and heavier rainfall.

In Bangladesh, the risk of flooding is predicted to rise by 20 percent in the next 20-50 years. Predicted yield changes for wheat, maize and rice by the year 2020 suggest that yields in Nigeria and Brazil will fall by 2.5-5 percent, and in India by 5-10 percent (although there are countries where yields may rise). Relatively small increases in temperature may spread malaria into large urban areas such as Nairobi and Harare that currently lie just outside the malaria range.

Source: IPCC (2001); IIASA (2001); CGIAR (2000).

Crisis and conflict

Tensions between diverse interest groups over natural resources can contribute to conflict, including in situations where resource mismanagement and environmental degradation intensify competition over access to shared resources. These tensions may be played out at a regional level, as can be seen in the water conflicts in the Middle East; at national level, for example the competition for control of diamonds in Sierra Leone; and at the local level over access to natural resources on which the poor directly depend for their livelihoods (DFID, 2000a). In such circumstances, the poor will be the most negatively affected because they have the least resources to cope with physical loss, and are the most vulnerable to violence and lack appropriate means for legal redress.

New research suggests that civil wars more often are fueled by rebel groups competing with national governments for control of diamonds, coffee, and other valuable primary commodities, than by political, ethnic or religious differences. Analysis of 47 civil wars from 1960-1999 shows that countries which earn around a quarter of their yearly GDP from the export of unprocessed commodities face a far higher likelihood of civil war than countries with more diversified economies. Since conflict prevention efforts have paid relatively little attention to these issues, there would seem to be considerable scope for both domestic and international policy to prevent civil conflict more effectively (World Bank, 2001a).

In some cases, natural resource conflicts can be so severe that they contribute to wider unrest and can affect the political stability of a country. In Burundi and Rwanda, there is some evidence that intense population pressure combined with limited land resources were contributing factors to the ethnic tension that led to full-scale civil war (ACTS, 2000). And there is evidence that some of the enduring conflicts in other African countries – for example in Angola, Democratic Republic of the Congo, Liberia and Sudan – have either arisen from competing desires to control rich natural resources, including conflict among elites over control of profits from natural resource exploitation, or have provided funds for the conflict to continue (ACTS, 2000; Global Witness, 2000 and 2001; Oxfam, 2002; Göeteborg University, 2002).

1.4 Economic growth and the environment

The links between growth, economic policies and the environment are important for poverty reduction in two inter-related ways:

☐ There is not a simple trade-off between growth and the environment – countries with similar levels of income and growth can have quite different levels of environmental performance;

☐ Ignoring the environmental soundness of growth – even if this leads to short-run economic gains – can undermine long-run growth and its effectiveness in reducing poverty.

The quality of growth matters

Current strategies for poverty alleviation are fundamentally built upon premises of economic growth. A wealth of empirical evidence reveals that economic growth, as commonly measured in increases of real Gross Domestic Product (GDP), is necessary but not sufficient to reduce the number of people living in poverty (World Bank, 2001). Economic growth is essential for poverty reduction, and so is its distribution.

Critical to discussing economic growth as it relates to environmental impact and poverty is the consideration of the quality of growth. The same rate of growth in the economy can be associated with widely different environmental impacts, as seen in Figure 3. Depicted on the y-axis are changes in environmental quality based upon a sustainability index measuring changes in water pollution and air pollution during the 1980s and deforestation over the 1980s and 1990s. The higher the position on the y-axis, the more a country's environmental quality ranking has improved. While this type of un-weighted, simple index only partially covers the concept of environmental quality, it serves to illustrate a fundamental point. We are not dealing with a simple trade-off between growth and environment since there is a wide range of environmental performance scores for a given level of GDP growth.

FIGURE 3: ECONOMIC GROWTH AND ENVIRONMENTAL QUALITY (1981-1998)

Source: World Bank (2000c)

As economies grow, their environmental performance tends to deteriorate or improve depending on what variable one considers. Comparing across countries at different income levels:

☐ Water quality tends to improve with rising income;

☐ Air pollution from sulfur dioxide tends to first get worse with rising income, but then decline;

☐ Finally, the emission of carbon dioxide tends to continue to grow with income, although not uniformly so (World Bank, 1992).

These are comparisons across country income groups, but countries at similar income and growth levels show large differences. The bottom line is simple; policy matters.

Ignoring the environment can undermine long-term growth

While there is no simple relationship between growth and environment, there are many examples of how bad environmental management is bad for growth. These short-run growth paths are bad for long-run growth, but also have high social and environmental costs. Some examples include:

☐ Collapse of fisheries in many countries both in the developed and developing world – for example, the cod fishery in the North Atlantic and various fisheries in the developing world.

☐ Decline of agriculture due to salinization from irrigation in several countries, for example in Pakistan. It is has been estimated that about 16% of the country is subject to salinization from low quality groundwater provided by tube wells and excessive water application. The damage from salinization costs the country over US$200 million per year in reduced yields (World Bank, 1996). Another example of unsustainable irrigation was the draining of the Aral Sea to grow cotton, which has cost the region millions of dollars.

☐ Downstream impacts due to upstream land use change. Understanding the linkages between land use and downstream siltation and flooding are complex – but there is some evidence of the links. For example, the Chinese government has concluded that the severe flooding of 1998 was caused in large measure by deforestation in the Yangtze River's watershed.

☐ Decline in exports of intensively-farmed commercial aquaculture operations, in particular shrimp farming due primarily to disease from pollution and poor environmental control. The Taiwanese shrimp industry collapsed after the introduction of diseased animals. Newly discovered viruses caused losses of over a billion dollars in Asia in the 1990s. Now the shrimp industry in Latin America is threatened by these same pathogens (Bartley, 1999).

PART 2
Policy Opportunities to Reduce Poverty and Improve the Environment

Part 2 looks at policy opportunities to improve the relationship between poverty, growth and the environment. Given the complex and multidimensional nature of poverty-environment linkages, it is inevitable that this discussion encompasses a very broad agenda for policy and institutional change. We have grouped these issues into four main areas of policy action:

FIGURE 4: POLICY OPTIONS TO ADDRESS POVERTY-ENVIRONMENT LINKAGES

☐ Integrate poverty-environment issues into national development frameworks;
☐ Engage civil society including poor and marginalized groups;
☐ Address gender dimensions of poverty-environment issues;
☐ Implement anti-corruption measures;
☐ Take steps to reduce environment-related conflict;
☐ Improve poverty-environment monitoring and assessment.
☐ Strengthen resource rights of the poor;
☐ Support decentralization and local institutional development for environmental management;
☐ Expand access to environmentally-sound and locally-appropriate technology;
☐ Promote measures that reduce the environmental vulnerability of the poor.

- ☐ Integrate poverty-environment issues into economic policy and decision-making;
- ☐ Improve environmental valuation;
- ☐ Expand private sector involvement in pro-poor environmental management;
- ☐ Implement pro-poor environmental fiscal reform.
- ☐ Reform trade and industrialized country subsidy policies;
- ☐ Make foreign direct investment more pro-poor and pro-environment;
- ☐ Enhance the contribution of multilateral environmental agreements to national development objectives;
- ☐ Encourage sustainable consumption and production;
- ☐ Enhance the effectiveness of development cooperation and debt relief.

The need for policy and institutional change

Experience demonstrates that, with judicious policy-making, significant 'win-win' opportunities exist to reduce poverty by improving the environment. If better environmental management can contribute to poverty reduction, how can these opportunities be taken and what is preventing their wider adoption?

Many of the underlying causes of poverty and environmental degradation are related to issues of governance and politics. There are significant and often deeply entrenched policy and institutional barriers – at local, national and global levels – that work against the interests of poor and marginalized groups, and often create incentives to cause or overlook damage to the environment.

The past decade of experience since the 1992 Earth Summit in Rio reveals some important lessons that help point the way forward. Four broad lessons are highlighted here:

☐ First and foremost, poor people must be seen as part of the solution – rather than part of the problem. Efforts to improve environmental management in ways that contribute to sustainable growth and poverty reduction must begin with the poor themselves. In many cases, policies continue to be based on uncertain assumptions and over-simplifications concerning the poor and their relationship to the environment. A considerable body of evidence now exists that supports an improved understanding of poverty-environment interactions, in particular how environmental conditions affect the poor and their access to environmental assets (see Box 5). Given the right incentives and support, the poor will invest in environmental improvements to enhance their livelihoods and well-being, by applying their own resources and knowledge and by adopting new technologies and practices that are appropriate to their needs and circumstances. At the same time, however, it is essential to address the activities of the non-poor since they are the source of most environmental damage.

☐ The spatial and temporal trade-offs and competing economic and political interests that often underlie environmental management decisions and practices need to be addressed in ways that benefit the poor. Developing countries can face difficult choices in allocating scarce resources among pressing development needs, and the environment is often viewed as a longer-term concern that must be traded-off to address short-term needs (as has often been the case in the industrialized countries). At the same time, many examples are known where efforts to protect the environment have not taken into account the interests of poor and marginalized groups and have left them worse off. There are rational ways of dealing with conflicting interests and trade-offs, but they require more participatory, transparent and accountable policy and decision-making processes to ensure their credibility and longer-term effectiveness.

☐ Ignoring the environmental soundness of growth – even if this leads to short-run economic gains – can undermine longer-run growth and its effectiveness in reducing poverty. The environmental quality of growth matters to the poor. Environmental improvement is not a luxury preoccupation that can wait until growth has alleviated income poverty, nor can it be assumed that growth itself will take care of environmental problems over the longer-term as a natural by-product of increasing affluence. First, this ignores the fundamental importance of environmental goods and services to the livelihoods and well-being of the rural and urban poor. Second, there are many examples of how bad environmental management is bad for growth, and of how the poor bear a disproportionate share of the costs of environmental degradation. To improve the environmental soundness of growth, economic policies and decision-making must better reflect the 'public goods' nature of many environmental goods and services by addressing the persistent policy and market failures that lead to their under-valuation.

☐ Environmental management cannot be treated separately from other development concerns, but requires integration into poverty reduction and sustainable development efforts. Improving environmental management in ways that benefit the poor requires policy and institutional changes that cut across sectors and lie mostly outside the control of environmental institutions – changes in governance, domestic economic policy, and in international policies.

BOX 5: COUNTERING SIMPLISTIC NOTIONS ABOUT POVERTY AND ENVIRONMENT

☐ Most environmental degradation is not caused by the poor:
Most environmental degradation is caused by the non-poor as a result of their production and consumption levels, which are much higher than those of the poor, particularly in the highly-industrialized countries. Even where poor people degrade the environment, this is often due to the poor being denied access and full resource rights by wealthier elites, or being hired to engage in environmentally-damaging practices.

☐ Population growth does not necessarily lead to increased degradation:
While increasing population undoubtedly places greater pressure on productive land and resources, it is not necessarily population per se that causes the damage. The complex of locally-specific social, economic, environmental and governance circumstances in which increasing population takes place – which in turn can be strongly influenced either positively or negatively by external policy and institutional factors – are the driving forces behind poverty-environment interactions. Indeed, conventional economic theory would suggest that as population increases and land becomes scarcer, the land should increase in value and merit greater care and investment. Research in Kenya has documented cases where, even in the face of increasing population pressures, farmers have managed semi-arid, degraded, unproductive lands in a manner that has rehabilitated them and made them profitable (Tiffen et al., 1994). A wider review shows that for population growth to lead to improved soil and water investments, market access and attractive producer prices are essential, as well as social and economic support to prevent the collapse of social structures (Boyd and Slaymaker, 2000).

☐ The poor will invest in environmental improvement:

The conventional wisdom has been that poor people are too impoverished to mobilize resources for enhancing the environment. In some cases this is true. But numerous experiences demonstrate that when incentives are favorable, low-income households and social groups can mobilize enormous resources, particularly labor. Many urban environmental problems can most effectively be solved when poor communities mobilize themselves or form coalitions with less poor groups to improve service provision, often with some contribution in cash or kind (Satterthwaite, 2001).

☐ Poor people often have the technical knowledge for resource management:
It is often assumed that a lack of technical knowledge is a key constraint to poor people's management of natural resources. Indeed, when poor people move to areas with new ecological regimes, or when something happens to change the balance under which their resource management practices developed, a period of adjustment is required. Evidence is increasingly showing that poor people have an enormous store of what is sometimes termed indigenous technical knowledge, such as use of medicinal plants, water harvesting structures, fishing sites and so on. This knowledge is often undervalued or completely ignored. There are many well-documented cases of poor people investing their own time and resources in environmental management, and succeeding in maintaining production and profitability, while keeping their families and communities from the worst effects of poverty.

2.1 Improving governance

Key areas for policy action:
☐ Integrate poverty-environment issues into national development frameworks
☐ Engage civil society including poor and marginalized groups
☐ Address gender dimensions of poverty-environment issues
☐ Implement anti-corruption measures
☐ Take steps to reduce environment-related conflict
☐ Improve poverty-environment monitoring and assessment

Poor people are very capable of sustaining and improving their own livelihoods as long as they have adequate opportunities and capabilities to make a living, and a voice in decisions that affect them. Although poverty reduction strategies and related country planning frameworks are intended to reflect the poor's priorities, issues that matter most to the poor, including poverty-environment links, often have been overlooked. This points to the need for open, transparent and accountable policy and decision-making processes that respond to the realities of the poor. Improved governance, including an active civil society, is often the missing link in creating a more enabling policy and institutional environment to address poverty-environment issues that matter to the poor.

Integrate poverty-environment issues into national development frameworks

Addressing governance issues at both national and sub-national levels is vital in promoting pro-poor policies and programmes, including the integration of poverty-environment issues into national planning processes for poverty reduction and sustainable development. Politicians, the judiciary and the civil service all have a role to play as the state directly controls access to many natural resources, or determines the rules for resource use, controls investments for environmental infrastructure and creates the framework for public policy debate about poverty-environment issues.

Poverty-environment issues need to be integrated into mainstream development planning and resource allocation processes – national development plans and budgets, poverty reduction strategies, and sector plans and budgets – in order to create an enabling policy environment for a broad-based and coordinated response. This is necessary to achieve synergy between diverse interventions across many sectors, and to ensure that adequate resources are being allocated from both domestic and external sources.

All countries have some form of national strategic planning process. At the 1995 World Summit for Social Development, governments committed themselves to developing more explicitly pro-poor policy frameworks through the preparation or strengthening of national strategies to reduce poverty. In 1999, the World Bank and the International Monetary Fund (IMF) made Poverty Reduction Strategy Papers (PRSPs) the basis for debt forgiveness and new concessional lending. Nationally-owned poverty reduction strategies, including the PRSP process, provide a critical entry point for integrating relevant poverty-environment issues and ways to tackle them into a country's mainstream development policy framework.

Recent environmental reviews of PRSPs prepared in 40 countries found a mixed picture (DFID, 2002b; Bojö and Reddy, 2002). Some countries, such as Bolivia, Honduras, Mozambique, Nicaragua and Uganda (see Box 6) have made a significant effort to address the issues of improved natural resource management, better environmental health and disaster preparedness. However, in most other countries, these issues have not been adequately addressed in the context of their poverty reduction planning. Even where environmental matters are adequately addressed in PRSPs, considerable work still needs to be done to ensure that Medium-Term Expenditure Plans and line ministry budgets contain adequate and properly directed resources for pro-poor investment in the environment.

At the 1992 UN Conference on Environment and Development (UNCED), governments made a commitment to adopting national strategies for sustainable development, and this commitment is reflected in the Millennium Development Goals (e.g., Goal 7 on "ensuring environmental sustainability"). The UN has prepared guidance to assist countries in preparing a sustainable development strategy (UNDESA, 2002), and the OECD Development Assistance Committee has prepared similar guidelines for development agency support to such processes (OECD, 2001). Each country needs to determine its own strategy process. The challenge is to seek convergence between poverty and sustainable development strategies, and avoid the continuing tendency of donors to promote multiple and competing strategy frameworks. Where Poverty Reduction Strategies adhere to their stated principles including the integration of relevant environmental issues, then this can be considered a national strategy for sustainable development (OECD, 2001; DFID, 2000c).

BOX 6: INTEGRATING ENVIRONMENT IN UGANDA'S POVERTY ERADICATION ACTION PLAN

In early 2000, Uganda's Poverty Eradication Action Plan (PEAP) was updated. Early drafts of the revision contained little recognition of environmental issues and long-term sustainability. For example, the focus in energy policy was on electrification, although fuelwood accounts for 96 percent of domestic energy supply. The National Environmental Management Authority (NEMA) engaged in the process by producing a series of amendments and additions that were incorporated into the strategy. Many of the inadequacies were simply due to oversight and were gladly incorporated when the merits were understood. Other parts of the Ministry of Water, Lands and Environment submitted their own PEAP amendments once the influence of the NEMA initiative became known. Since the PEAP was adopted, NEMA has been engaged in following up sectoral plans for example, the Plan for the Modernization of Agriculture and in identifying poverty-environment indicators to ensure that poverty-environment issues are implemented.

Source: DFID (2000b).

Environmental issues that matter to the poor need to be fully integrated into sectoral plans and policies. Promoting commercial farming that drains wetlands without thinking of the impact this will have on existing users of the wetland is short-sighted and may negatively impact the poor. Promoting an energy policy that focuses only on electrification, which the poor cannot afford and so will remain dependent on fuelwood, is counter-productive. Funding more rural health clinics, without investments in environmental health, is also not cost-effective. All policies need to be assessed to ensure that environmental opportunities to help the poor have not been overlooked (Yaron and White, 2002).

A shift to more integrated and cross-sectoral approaches to addressing poverty-environment issues does not imply a less significant role for Environment Ministries and natural resource-related agencies, nor does it reduce the need for adequate funding, staffing and training to carry out their policy and regulatory mandates. However, it does mean that environmental organizations – including in civil society – need to better understand how environmental conditions impact the poor and the ways in which environmental management can contribute to poverty reduction. It also means that environmental organizations must engage more effectively with Ministries of Planning and Finance, or other agencies driving the national planning process, to ensure that poverty-environment issues are addressed. In most cases, this shift in orientation will require a reassessment of environmental mandates, legal and regulatory frameworks and capacity development needs.

Engage civil society including poor and marginalized groups

The participation of poor and marginalized groups in policy and planning processes is essential to ensuring that the key environmental issues that affect them are adequately addressed. It also fosters commitment to the environmental policies and interventions to be implemented. The effective participation of these groups depends on a number of factors. The participatory mechanisms put in place should be sensitive to the resource constraints of poor people, particularly their time. And they must enhance transparency and accountability in order to convince poor people that their views will be considered and given due weight.

Where government is responsive, it can have a major impact. In India, reformist governments in the states of West Bengal and Andhra Pradesh were instrumental in promoting greater joint management by the poor of forestry resources (Lele, 2000). In several Latin American cities, progressive mayors and city councils have made a major impact in improving the access of the poor to environmental infrastructure (Hardoy et al, 2001).

Access to information is vital for effective environmental management. A free media has been instrumental in highlighting environmental problems in both the public and private sectors. In some countries, the state has effectively used public pressure by making information publicly available in order to encourage greater pollution compliance (see Box 7). This also applies to rural areas. In the Philippines, for example, access to information has contributed to community monitoring of forestry offences and the enforcement of forest regulations (Brunner et al., 2000).

BOX 7: INDONESIA'S PROGRAM FOR POLLUTION CONTROL, EVALUATION AND RATING (PROPER)

The Indonesian environment agency, BAPEDAL introduced PROPER in early 1995 focusing on 187 of the worst water polluters. The Vice President presided over a high profile ceremony to congratulate the one third of companies that met the regulations, while BAPEDAL privately notified the remaining two thirds that they were non-compliant and had six months to go before public disclosure. Following full disclosure, the program had by mid-1997 reduced pollution by 40%. Indonesia is now expanding the program to 2000 plants. Other countries have learnt from this approach and similar schemes are now underway in the Philippines, Mexico, and Colombia, and are planned in China and Venezuela.

Source: World Bank (2000b).

Address gender dimensions of poverty-environment issues

Gender-related issues are a key dimension of the poverty-environment nexus (OECD, 2001). Low investment in female education reduces a country's productivity. Rigid gender roles contribute to inefficiencies in natural resource management. As described in Part 2, women are at higher risk and more vulnerable than men to many environmental hazards because of their particular social and economic roles. To date, poverty-environment links that matter to poor women – such as lack of land and resource rights, the additional disease burden from indoor air pollution and the time and physical burden of collecting fuelwood and water – have been given very little recognition in almost all PRSPs. Existing gender analysis methods and tools must be employed to ensure that poverty reduction strategies reflect a more gender-disaggregated understanding of poverty and environment concerns.

Implement anti-corruption measures

Corruption is a general governance problem, but relates strongly to poor environmental management, especially concerning the extraction of natural resources, the regulation of pollution and the preference for lucrative hardware solutions (for example, the power and water sectors) over softer solutions like efficiency savings. The Environmental Sustainability Index (ESI) found that the variable that most correlated with poor environmental performance was corruption.

The provision and effective dissemination of good quality information, combined with an appropriate legal and regulatory framework and the eventual imposition of adequate sanctions, can improve the situation. Pressure can be brought to bear by national and international civil society, by international buyers and consumers, by donors, and by other governments (see Box 8). For example, according to Article 97 of the Cotonou Agreement between the European Union and ACP (African, Caribbean and Pacific) countries, serious cases of corruption should give rise to consultations between the Parties to the Agreement, and require the Party where the serious cases of corruption have occurred to take the measures necessary to remedy the situation immediately. In some cases, sanctions may be imposed such as suspension of aid.

While developing countries have a role to play in stamping out corruption, developed countries also can play a part – as they may be home to the briber. Recently the OECD passed a Bribery Convention – which says member states should make it illegal to bribe non-OECD national. The OECD requires government to introduce legislation to achieve this – which many OECD countries, such as the US and UK have done. There is also a desire by some developed country governments and businesses to agree multilateral rules to make it a requirement to make public the amount of rent taxes they are handing over to developing governments for legal exploitation – often for oil – but to ensure that this money does not disappear.

BOX 8: TACKLING CORRUPTION IN THE CAMBODIAN FORESTRY SECTOR

Cambodia's Interim Poverty Reduction Strategy states: "controlling illegal logging, combined with measures already taken to restructure the forestry concession system, will begin to mobilize the revenue potential of the forestry sector which will become an important source of finance for poverty reduction measures in agricultural and other sectors." It is estimated that about US$100 million are lost each year from corruption, compared to only about US$13 million that are captured. The Forest Crime Unit supported by the international NGO Global Witness has been very blunt about drawing attention to the lack of action against illegal loggers. Faced with mounting domestic and international criticism, Cambodian Prime Minister Hun Sen announced the suspension of all logging operations effective January 2002.

Source: Hodess (2001).

Take steps to reduce environment-related conflict

Environmental conflict is an issue at micro and meso level (e.g., pastoralists versus settled farmers, river basin users) and at a macro level (e.g., control of diamonds and timber fueling conflict). At the micro and meso level, conflict resolution structures are needed that provide a forum for informed dialogue to solve problems. For example, river basin management authorities are being set up in many countries to achieve dialogue and management rules between different river users. In some cases, the open access nature of many resources – land, fisheries, forests – needs to be altered to stop over-use which creates conflict. Local-level schemes to define management regimes need to be supported. This can be complex as it is important not to exclude poor people. For example, while protected areas are being managed with more involvement of local people – there are still too many examples of protected areas that are detrimental to poor people's livelihoods and well-being.

Control over natural resource rents – particularly oil and other minerals – can cause conflict between local residents, governments and private extractors. In some cases, there also can be tension between the local district where the minerals are located and central government – who may get much of the revenue – an issue that has arisen in Indonesia, Papua New Guinea and Nigeria. This can be addressed by reaching a political settlement on the appropriate and transparent sharing of resource revenues.

In more extreme cases, natural resources may fuel war. While natural resources rarely cause conflicts, they often provide the funds and incentives to prolong conflicts once they have started. This has been the case in West Africa and South East Asia. The underlying cause for the conflict needs to be addressed, but in the meantime pressure from the international community – governments, civil society and consumers – can reduce the potential gains from resource extraction. The Kimberley diamond certification process is one such attempt, as is pressure by the UN Security Council to highlight natural resource extraction in the Democratic Republic of the Congo (see Box 9).

BOX 9: NATURAL RESOURCES FUEL CONFLICT IN THE DEMOCRATIC REPUBLIC OF THE CONGO

In the Democratic Republic of the Congo (DRC), the link between conflict and natural resources is now so explicit, the United Nations Security Council asked the UN Secretary general to set up in 2001 a special expert panel on the illegal exploitation of natural resources and other forms of wealth in the DRC. The Panel argued in their first report that there is "a pattern of continued exploitation carried out by numerous state and non-state actors, including rebel forces and armed groups, conducted behind various facades in order to conceal the true nature of the activities". The only loser in this huge business venture is the Congolese people. Following a debate on the panel's conclusions in December 2001 its mandate was extended to include an update of information from all relevant countries; an evaluation of possible actions that could be taken by the Security Council in order to help end plundering; recommendations on specific actions that the international community might take in support of the Congolese government; and recommendations on possible steps that might be taken by transit countries, as well as end users, to contribute to ending illegal exploitation of natural resources.

Source: IRINnews.org, UN Office for the Coordination of Humanitarian Affairs, 2002

Improve poverty-environment monitoring and assessment

To assess how better environmental management can reduce poverty requires local understanding of how environmental conditions relate to dimensions of poverty, and the ability to identify and prioritize alternative policy options and to evaluate their effectiveness and impact. This, in turn, requires appropriate and effective indicators and monitoring systems. Environmental data tend to focus on environmental change without determining poverty effects, while poverty monitoring systems often ignore environmental concerns. Indicators are needed that measure how environmental conditions affect the livelihoods, health and vulnerability of the poor, and these indicators need to be integrated into national poverty monitoring systems and assessment.

Some work is already underway to identify useful generic poverty-environment indicators, but the real need is to collect data in country. Surveys in Nepal,

Honduras and Uganda (Nunnan et al, 2001) and Nigeria (Osuntogun, 2001) show that some data is already available. Generally, environmental health data are currently the most widely available, drawing from Ministry of Health and household survey sources. However, the extent to which certain health outcomes such as malaria can be reduced by environmental interventions requires further research. There are some qualitative data on natural resources and vulnerability from participatory poverty assessments (PPAs), but future PPAs could be designed with a more explicit focus on key poverty-environment issues (Brocklesby and Hinshelwood, 2001). Micro-level data on the poor's dependence on natural resources are sometimes available for a particular sector, such as the forestry sector, often as part of preparing forestry sector and biodiversity strategies. Work has also been undertaken to overlay poverty data with existing environmental data to form "poverty-environment maps" that identify the spatial links between poverty and resource degradation (Henninger and Hammond, 2000). While this suggests that data may be more available than is realized, it is scattered among different agencies, not collected systematically and often requires careful analysis and interpretation to develop its relevance for poverty-environment issues.

As with any indicators, the information collected is only useful to the extent that it is appropriately used. Poverty-environment data collection should build on existing data collection efforts such as those associated with livelihood surveys and participatory poverty assessments, and be anchored in institutions with appropriate skills such as the Statistics Department, Ministry of Finance or a competent local research institute. These institutions have experience in producing demand-led data and will make it more likely that the data is fed into ongoing poverty-related policy processes such as poverty reduction strategies and sectoral and spatial plans and programmes.

2.2 Enhancing the assets of the poor

Key areas for policy action:
- ☐ Strengthen resource rights of the poor
- ☐ Support decentralization and local institutional development for environmental management
- ☐ Expand access to environmentally-sound and locally-appropriate technology
- ☐ Promote measures that reduce the environmental vulnerability of the poor

In "mainstream" development, there is a growing recognition that poor people have the capacities to lift themselves out of poverty if they have the right support and incentives. This is especially important in the domain of poverty and the environment, where poor people increasingly are the stewards of the environment that sustains their livelihoods and well-being. In the poorest countries where natural resources represent the most obvious source of wealth, their control is about power. In cases where existing institutions and environmental management arrangements generally favor the poor, they should not be undermined by the state. However, the state may have an important role to play in cases where traditional arrangements are breaking down under internal or external pressures.

Strengthen resource rights of the poor

Property rights to resources such as land, water and trees have been found to play a fundamental role in the poverty-environment nexus (Scherr, 1999). Property rights encompass a diverse set of tenure rules and other aspects of resource access and use, and strongly influence the patterns of natural resource management. They may either facilitate or impede sustainable use, protection or resource-improving investment.

Property rights held by poor people represent key household and community assets that may provide income opportunities, the ability to meet essential household subsistence needs, and/or provide a means of insurance against livelihood risk. Poorer people tend to rely more heavily on customary or informal rights that are not adhered to by outside user groups. Marginalized users, such as poor women, often lose out as a result of policies and processes which privatize and reduce complex bundles of rights into a single unitary right (under many land and water reforms). Uncertain ownership conditions can also affect long-term agricultural productivity and incentives for resource conservation and investment, and can especially cause rapid deterioration of lands or natural resources when the owner tries to squeeze out the maximum revenue during a short period.

Good examples are available of well-established common-property management regimes that do not meet the criterion of private exclusivity, and yet function to the satisfaction of the included parties and have proven to be sustainable (Ostrom, 1990). There are also strong concerns that a shift toward privatization would be contrary to poverty alleviation: the rich tend to be the largest landowners after common land is privatized. Where traditional common property management regimes have broken down, the formal issuance of legal titles may be beneficial for agricultural productivity, and therefore create an incentive for investment in soil and water conservation (see Box 10). However, as perceived security and local enforcement are critical concerns, such formal titling may not be necessary if informal rules of the game are honored.

To strengthen the land rights of the poor, it is necessary to reform the policies and institutions responsible for delivering land rights in order to make them more responsive to the poor's needs. These include central government land agencies, local government, traditional authorities, the justice system, and local land boards, commissions and tribunals.

BOX 10: LAND TENURE AND ENVIRONMENTAL IMPROVEMENTS

The relationship between land tenure and environmental improvements in terms of afforestation and soil and water management in rural areas, and investing in better housing in urban areas, are complex and location-specific.

A study of 115 upland farms in the Philippines using 6 years of soil erosion data found that farmers who had high security of tenure were more likely to install contour hedgerows to reduce erosion (Shively). However the study also found that adaptation is more likely with farms that have access to credit, and that larger farms are more likely to adapt than smaller farmers. This suggests that while tenure is important it is by no means the only factor that matters.

However studies from parts of Africa are less categorical – showing that while tenure is important, tenure security is not necessarily delivered by freehold titling (DFID). Tenure security is often a question of perception and interpretation of the socio-political climate in relation to land rights.

Source: Shively, G (2001); DFID (2002a).

Support decentralization and local institutional development for environmental management

With the trend toward greater decentralization and devolution in many countries, planning is increasingly being undertaken at a state, district or local level. For example, many countries such as Tanzania, Egypt and Sri Lanka have introduced district-level environmental planning. While this is an important development, it is vital that these environmental plans are integrated into the mainstream local planning process. It is also important that these plans focus on issues which are relevant to poor people – approaching the issues from their perspective, and not only from an environmental perspective.

Decentralization in rural areas has given local governments control over many key natural resources – such as state land – and responsibility for infrastructure such as water supply, sanitation and irrigation. Rules on resource access – such as permits for mining, timber harvesting, grazing and industrial emissions – are generally issued by local government. In cities, up to half of urban land is commonly in the public domain as public buildings, public infrastructure and land (e.g., roads, railways, canals). The way local government chooses to use this land affects where industry locates, how congested a city is, where people live and how the city will develop (DFID, 2001b). Further, urban environmental problems can most effectively be solved when poor communities are able to mobilize themselves or form coalitions with less poor groups to improve service provision, often with some contribution in cash or kind (Satterthwaite, 2001).

While greater local government control has in some cases made decision-making more responsive and accountable, this is by no means guaranteed. Local governments can be subject to just the same "capture" by wealthy elites as central government, they can also manage local resources unsustainably to raise revenue and may have weaker capacity than central governments. In addition, decentralization has often been undermined when central governments have not provided sufficient resources transfers or revenue raising powers for local government to implement their responsibilities.

To counter these problems, it is essential to build social capital through decentralization and empowerment of groups within communities that favor sound environmental management, and supporting institutions that are honest, transparent, and have the confidence of the local population. Many positive examples of local empowerment can be cited (see Box 11):

☐ Community wildlife preserves managed for sport hunting in southern Africa have been transformed into protected areas established co-terminously with indigenous territories, where indigenous people's livelihoods become a force for conservation.

☐ Water-user's associations that buy and sell water rights and organize for collective system maintenance have been established.

☐ Community forestry enterprises linked to international timber and certification markets;

☐ Cooperatives producing organic foods or coffee, which revitalize traditional agricultural systems with new technologies.

In all of these examples, the institutional framework, including the building and use of social capital, is a key element in success. Projects that successfully support such initiatives have included significant resources for human capital development, organizational strengthening, negotiation and conflict resolution, and other institutional skills. Community-level organizations have also developed relationships with higher-level institutions, and through them mobilized support for their interests and advocated a positive policy environment for their activities.

BOX 11: COMMUNITY FORESTRY IN NEPAL

The 1993 Forest Act legalized forestry user groups giving them the right to own the trees, although ownership of the land remains with the State. User groups develop operational plans, set forest product sale prices, and determine how surplus income is spent. By June 1997, there were 6,000 user groups managing 450,000 ha with a further 6,000 waiting for formal registration. Issues still arise within user groups, between them, and with the forest department. Concerns have arisen about domination by local elites, politicization of user groups, and pressures from the forest department to focus on tree planting rather than harvesting. Nevertheless, experience has been encouraging; and the condition of the managed forests has often improved.

Source: Arnold and Bird (1999).

Decentralization and local empowerment is not a guarantee for environmental stewardship. Not all stakeholders have compatible objectives and there are different degrees of power and influence. This can lead to conflicts when certain groups are left out of the process or when success encourages others to enter. Hence, efforts to empower communities to locally manage natural resources must safeguard against elite capture, and build local capacity for participatory management. Also, devolution of power to the local level can increase pressure on natural resources in view of the income, employment, and revenue needs of local government and their constituents. Hence, when tradeoffs between environmental conservation and poverty reduction are resolved locally, they may result in short-term exploitation. However, this can be mitigated by two factors. The first is that local resource control also means that the benefits of sustainable management will accrue locally. The second mitigating factor is that financial transfers from the outside, for example through nationally-directed subsidies or international funding sources such as the Global Environment Facility, can make a big difference as to how these tradeoffs are resolved.

Expand access to environmentally-sound and locally-appropriate technology

There is an abundance of "appropriate" technologies that can improve the environment and the livelihoods of the poor. Many are based on local traditional knowledge and practices, others are the result of external technical innovation. Examples include terracing, tied ridging to hold rain water, planting grass bunds to reduce water run-off and soil erosion, water harvesting techniques, agro-forestry, the use of natural products to eliminate pests, improved livestock and fish production, the use of reeds or woody plants to trap and detoxify sewage and many others.

However, technology development and dissemination for the poor is often not fully provided by the market. Because of its possible spill-over benefits, governments, civil society groups, the poor themselves and donors have a role to play to support innovation. Such shifts might be brought about through introduction and demonstration projects that involve the full participation of poor people. There have been attempts to fund labor-intensive environmental technology projects through public works, especially "food for work" programmes. However, the ownership and ultimate sustainability of works that have been carried out with the incentive of an external supply of income is usually questionable.

Much more success has been achieved by empowering innovative poor people to adopt and adapt new technologies and to pass their knowledge on to their peers (Reij and Waters-Beyer, 2001). Support should be provided to involve farmers in testing the suitability of these new practices and the use of "farmer-to-farmer" advisory and training services, leading to the introduction of a number of different practices which require little or no cash inputs - a very important feature when dealing with poor farmers. The practices can be based, for example, on making the best use of the rainfall and waste products like animal manure and crop residues and whatever other organic material can be found on the farm.

In the area of human health, there is tremendous need for improved cook-stove technology to reduce indoor air pollution and associated acute respiratory infections. In the past, many such programmes have failed, but there have been countries where, especially in urban markets, the new technology has successfully taken off. The issue here as with all technology is to focus not just on the engineering side, but on the social, cultural, financial and marketing aspects of technical change.

Promote measures that reduce the environmental vulnerability of the poor

The poor have many informal mechanisms to manage the risks that they face every day. These include ways to reduce risk (e.g., use of common property resources, temporary migration), mitigate risk (e.g., income diversification and informal insurance through sharecropping, marriage and rotating savings schemes) and cope with shocks once they occur (e.g., sale of assets, reduced consumption, loans). These risk management strategies may be at the individual, household or more collective level (World Bank, 2001).

State attempts to reduce the vulnerability of the poor to natural disasters should strike a balance between measures designed to prevent shocks that will adversely impact the poor, and ex-ante measures that reduce the impact of such shocks on poor and vulnerable groups or enhance their ability to cope. Intervention strategies need to be based on the realities of the poor and the kind of environmental risks they face. Government attempts to improve storm-water drainage in the slums of Indore, India involved covering the drains, which prevented residents from being able to predict the severity of the flood as they previously did; the drains are more easily blocked by rubbish and can no longer be used to wash away excreta – thus the residents preferred the old system (WRI, 1996). In many environmental disasters, the majority of fatalities occur in the first 24 hours – long before national and international agencies arrive on the scene. So engaging local residents in disaster preparedness, mitigation and coping strategies is the only practical solution.

While natural hazards in general cannot be prevented, their impacts, and sometimes their magnitude, can be managed. There are four main approaches (ICRC, 2001):

☐ Reduce the causes of environmental vulnerability through measures addressed elsewhere in this report. For example, floods are strongly influenced by land and water management in upper catchments of watersheds. Good land-use planning and zoning can prevent a natural cycle of water flows from becoming a catastrophe. Fire breaks and early response can to some extent prevent wildfires to spread.

☐ Focus more on participatory risk reduction and mitigation. Early warning systems that effectively provide local people with adequate information to minimize impacts can be very effective. Building codes for houses and other infrastructure can ensure that they are equipped to withstand natural hazards to a reasonable degree. Countries that have taken this approach have made a major impact. In Bangladesh, following the 1991 cyclone when 140,000 people died, a major effort was put into local-level disaster preparedness and since then fatalities have dropped substantially – although thousands are still made homeless. Even in the terrible 1999 Orissa super-cyclone – when an estimated

10,000 to 40,000 people died – an additional 40,000 were saved by locally constructed and managed shelters.

☐ After disasters have happened, improve relief coordination and place a greater focus on building up resilient sustainable livelihoods with a focus on disaster preparedness. Relief coordination remains very poor with too little involvement of well informed groups on the ground – and there is still too much flying in of foreign supplies and experts, which often distracts from more pressing issues. Evidence shows that economic vulnerability is more of a constraint than environmental vulnerability – so there must be focus on long-term improvements through, for example, the introduction of more income-earning opportunities. This is constrained by both government and development agencies who still tend to separate disaster relief from long term development – so that relief is not sufficiently development orientated and development does not fully incorporate disaster mitigation.

☐ Ensure that funds are available for dealing with disasters. While the international community may provide some funds, countries may find it more predictable to set up their own contingency reserves. A number of countries in Latin America have already begun this process.

Once a disaster has struck, emergency response management and delivery of rapid support to affected areas is critical to bring down human losses. Economic recovery requires a well-managed response with quick-disbursing funds for clearing of disturbed sites, reconstruction, re-seeding of damaged crop land, micro-credit for commercial activities and so forth.

Addressing chronic long-term environmental vulnerability is even more complex and, as it is less visible, receives much less attention. Long-term solutions require addressing the reasons for environmental decline. In the short term, the key is to understand the poor's own coping strategies and motives. In rural areas, coping strategies of the poor may include reducing dependence on declining natural resources, through shifting to off-farm employment and in some cases migration. In urban areas, there is some evidence that the poor make a trade-off to accept certain environmental hazards – such as polluted slums – in order to improve their economic opportunities (WRI, 1996).

2.3 Improving the quality of growth

Key areas for policy action:

☐ Integrate poverty-environment issues into economic policy and decision-making
☐ Improve environmental valuation
☐ Expand private sector involvement in pro-poor environmental management
☐ Implement pro-poor environmental fiscal reform

Environmental problems often arise because effective market mechanisms do not exist or are insufficient. Hence, there is an important role for government to complement structural reforms with measures to promote pro-poor environmental management. This includes the need to make environmental information more widely available to civil society and the private sector, and to better account for the economic values of environmental good and services ignored by markets, in order to make rational and enlightened choices possible. However, it is also important that governments correct the failures of their own polices. This refers to reform of perverse subsides, combating corruption and providing an enabling environment for the private sector to manage environmental resources when this can be done efficiently and in the best social interest.

Integrate poverty-environment issues into economic policy and decision-making

To promote macro-economic stability and enhance growth, many countries have undergone structural adjustment reforms that include exchange rate re-alignment, public sector reform and privatization, reduction of tariffs and subsidy reform. The effect of these past reforms on the environment is controversial and mixed. Positive environmental impacts can occur when, for example, an over-valued currency is adjusted so that domestic nature-based tourist services are promoted, or when public subsidies to polluting industries are dismantled. Adverse environmental effects can occur when these reforms are undertaken in the context of unchanged institutional and market failures. Trade liberalization can enhance export opportunities for natural resources such as forestry, fisheries and minerals. However, if these resources are open-access and environmental regulation and management regimes ineffective, the repercussions may be quite negative from both an environmental and poverty reduction perspective.

Many countries have had to adjust unsustainable economic polices, but there is a need to complement such adjustment in two important ways. First, economic policy reforms need to be complemented with assessments of their poverty-environment impacts. Traditional environmental impact assessment is now being adapted to address economic policy changes. Strategic Environmental Assessment (SEA) can be applied to sectoral and regional programs to identify potential impacts and design mitigating measures. Major deficiencies in environmental management can be identified and mitigation can be designed. For very broad macroeconomic reforms, however, it becomes difficult to predict what the ultimate impact on the environment will be. As numerous case studies have shown, the impacts often can be traced through chains of both positive and negative repercussions, but quantifying the impacts remains extremely difficult. Even after the implementation of an economic adjustment program, it remains a challenge to define the "without scenario", that is, what would have happened in the absence of the reform program.

Traditional economic models can have environmental components included – for example, to find out the effect of timber trade liberalization on forest cover. However, traditional economic and environmental analyses both need to be adapted to our current concern: that those impacts which affect the poor more are given greater attention. In some cases, countries are already starting to experiment with Poverty and Social Impact Assessment (PSIA) of policy changes, and there is a need to ensure that relevant poverty-environment issues are also captured.

This brings us to the second important point: there is no substitute for targeted support to environmental management capacity in a reforming country. While not every impact of reform can be foreseen in advance, certain environmental standards and monitoring capabilities can respond to and mitigate negative impacts that occur.

Improve environmental valuation

Markets form the backbone of the global economic system, but they fail to capture many important environmental values. This warrants attention both at the macroeconomic level, where social planning occurs, and at the microeconomic level, where households and individuals make small everyday decisions that, taken together, profoundly affect the level of environmental quality.

To make rational choices when environmental and economic values are to be compared, it is essential that accounting systems and market prices reflect the relevant values. At the macroeconomic level, this means that the traditional system of national accounting needs to be amended to better reflect environmental values.

Two main types of amendments are needed from an environmental perspective. First, the national income accounting system needs to differentiate between income derived from sustainable use of resources, and income derived from liquidation of natural capital. Second, water, soil and air pollution affect the level of environmental quality, and sometimes the productive capacity of the economy directly. In the latter case, the traditional income account already incorporates the negative impact of pollution. While no further adjustment to income is necessary, it is still of policy-relevance to trace the magnitude of the impacts. However, in the case where pollution does not directly affect current productivity, but non-marketed environmental services, or future productivity by inflicting long-term health damage, an amendment in the national income accounting is needed to reflect this.

The policy signals emerging from the national accounting data can be quite different if adjustments for subtractions/additions of human and natural capital are taken into consideration. One method is to derive an adjusted measure called Comprehensive Savings. Starting with the standard concept of net domestic savings, the current expenditures on education are added as an approximation of investment in human capital. Next, the depletion of non-renewable energy sources, minerals, and forests, are deducted. Finally, the damages from carbon dioxide emissions (as a proxy for overall air pollution) are deducted (World Bank, 2001e). This is illustrated in Figure 4, which shows a pronounced difference between the net domestic savings measure and the calculation of comprehensive savings for Sub-Saharan Africa. From a poverty reduction perspective, this type of macro-level analysis needs to be complemented with a distributional analysis – how does natural resource exploitation and ecosystem change, pollution and investment in human capital affect the poor?

FIGURE 5: ADJUSTED NATIONAL INCOME ACCOUNTS IN SUB-SAHARAN AFRICA

Source: World Bank (2002c)

Moving on from the perspective of society as a whole and down to the micro-level of individual and household decisions, poor people, like everyone else, will be influenced considerably by market prices. If market prices for environmental goods and services are not available, they need to be derived, using techniques of environmental economics. In summary, the incentives for people to make rational choices need to be improved. This is borne out in an example from Cambodia, where it was shown that local fisheries were damaged by the destruction of mangroves to make room for shrimp farms. Furthermore, the shrimp farms polluted the water, which further brought down catches for the traditional fishermen. The economic analysis showed that local communities in general would benefit from conserving the mangroves (Bann, 1997). Results from environmental economic analysis must be translated into policy and implemented if they are to have an impact on people's actions. This could be done, for example, through imposing fees on the harmful activities (clearing of mangroves, establishment of shrimp farms). This will not only discourage such activities, but it may also be a vehicle to compensate those who suffer the consequences.

Expand private sector involvement in pro-poor environmental management

With increasing liberalization in many countries, the role of the private sector has expanded, and the private sector is now an important player in terms of its ability to implement sustainable practices and as a source of expertise and funding. The impacts on poverty-environment issues are mixed, but are heavily dependent on the way the private sector is both managed and regulated (see Box 12).

Governments need to maximize the efficiency gains from the private sector while safeguarding the interests of the poor. For example, while privatization can improve the economic efficiency of environmental services such as waste management, sanitation and wastewater treatment, governments may need to provide safeguards to ensure that access by the poor is protected and improved. At the same time, governments need to increase their capacity for environmental regulation of private sector operations and enforcement of compliance. Particular attention should be given to ensuring that private sector operators integrate environmental considerations into their operations. This can include promotion of environmental management systems, environmental auditing and reporting and adherence to internationally-agreed codes of conduct.

BOX 12: THE 'FORESTRY PARTNERS PROGRAM' OF ARACRUZ CELULOSE IN BRAZIL

Aracruz Celulose operates in the southern part of the Brazilian state of Espírito Santo. This region has been under considerable pressure as a result of successive cycles of logging, agriculture, cattle raising and charcoal burning. The loss of native forest area has also been exacerbated due to the high levels of timber extraction by poor local communities.

Aracruz pulp has always been produced from planted forestry and avoided the exploitation of native trees for pulp manufacture or any other use. Under the Aracruz Forestry Partners Program, partnerships are formed with local farmers to develop new, sustainable timber plantations that provide alternative planted sources of timber for the company's pulp mill, and a new source of income to the farmers and local communities. Seedlings of native tree species are also provided for use in protected reserves. Moreover, guidance on environmentally sound techniques for managing plantations is provided to all farmers.

By mid-2001, the timber from the program was already responsible for 20% of the wood consumed by the Aracruz pulp mill. But more than this, it has played a significant role in improving the livelihoods of local farmers and their communities, as well as addressing important biodiversity issues. Participating farmers have generally been able to plant their eucalyptus trees on degraded or fallow land that is unsuitable for other locally grown cash crops; and as a result they have an additional source of farm income.

However, full privatization of environmental services may not be desirable or possible. A private company may not find it profitable to invest in potable water or sewage services for the poor, and strong trade unions may oppose private sector involvement if they fear heavy job losses. A promising approach to bringing in private sector investment is the establishment of public-private partnerships. These are arrangements where a government (national or local) enters into an agreement with a private sector enterprise to deliver investment and services within a jointly-agreed regulatory framework that safeguards the interests of the population to be served. Public-private partnerships are an increasingly common approach to expanding and seeking to improve environmental services such as potable water supply, sewage services, efficient transport and efficient energy production.

There has been an increase in private sector participation in the water services sector (water supply, irrigation and hydropower) in recent years (see Box 13). Still, the share of private sector water services is only about five percent worldwide (World Bank, 2002). The impact is the subject of a major controversy. However, experience to date indicates that public authorities will need to ensure that the service providers do not use their market power to exploit customers and that they internalize public health and environmental externalities. Public authorities also must ensure that water consumption is at a sustainable level, provide mechanisms to ensure that water supplies are efficiently allocated between alternative uses, and serve as a guarantor of a level of service provision that is consistent with a basic standard of living (Johnstone et al., 1999).

BOX 13: PUBLIC-PRIVATE PARTNERSHIPS FOR WATER SERVICES IN SOUTH AFRICA

In 1994, South Africa's first post-apartheid government produced a policy paper on Community Water Supply and Sanitation and, in 1998, passed the National Water Act. South Africa is an example of an enabling legal framework which opens concrete and well defined spaces for local action in terms of decentralization of powers, rights and responsibilities to the local level, as well as guidelines and regulations to help promote social equity and environmental sustainability.

This created a major capacity development challenge that has led to experimentation with public-private partnerships for developing water systems for the poor. The government funds construction, while users must pay for maintenance through a fee-based system for water services. The construction is done by private contractors who work for the users, who are organized as a for-profit organization. A European Commission review in 1999 found that this approach had provided 5 million people with water, completed 205 water projects and created 310,000 jobs. While the scheme has not been without problems, it has demonstrated the potential for scaling-up successful community-level water systems through innovative collaborations between government, the private sector, civil society organizations and the users themselves.
Source: Waddell (2000).

Implement pro-poor environmental fiscal reform

Subsidies. Environmentally-harmful subsidies are a key area for policy reform. These are subsidies that are both financially quite costly and lead to the overuse of natural resources and other unintended side effects, such as increased pollution. It is important to acknowledge that the largest such subsidies are handed out in industrialized countries (see Section 2.4).

Environmentally-harmful subsidies also are common in developing countries, particularly in the agriculture and infrastructure sectors. While many subsidies have been reduced or eliminated as part of structural adjustment and other policy reform processes (see Box 14) – for example, the removal of pesticide subsidies in Indonesia – the under-pricing of natural resources such as water for irrigation and various forms of fossil energy continues in many countries. Cost-recovery for irrigation water is only 10-25 percent in some of the major developing countries. Subsidies to gasoline and diesel in developing countries are in the order of $13 billion, and subsidies to electricity amount to more than $100 billion (World Bank, IMF and UNEP, 2002).

Subsidies to electricity can also be environmentally beneficial, as it encourages substitution from dirty fuels. However, these subsidies are often regressive as the rich benefit much more than the poor – for example, the poorest often are not served by subsidized electricity, water and waste collection. Even where the poor do get some benefit, subsidy reform can be structured to significantly increase its 'pro-poor' effect and to be less environmentally damaging. For example, tariffs for water or electricity can be differentiated to ensure the poor a basic supply at a "lifeline" rate, while raising the marginal cost for large-volume consumers. There are other ways to directly target the poor to raise the standard of living in general, without subsidizing specific commodities that the rich also consume. The potential impacts on the poor and the environment of alternative approaches to subsidy reform should be reviewed through environmental and social impact assessments and be subject to public comment before they are adopted.

BOX 14: ENERGY SUBSIDY REFORM AND THE POOR IN CHINA

China has made major strides in reforming its energy subsidies, particularly those to the coal industry, with significant benefits in terms of reduced pollution. Total economic subsidies for fossil fuels fell from $25 billion in 1990/1 to $10 billion in 1995/6. However the remaining subsidies still benefit the wealthier households, as most of the subsidized coal goes to urban areas. In rural areas, households depend on biomass and coal for cooking bought on the free market. Even where subsidized coal is distributed in rural areas, such as Western Xiushui, it primarily benefits higher income households. Rural energy is also consumed by Town and Village Enterprises (TVEs), but where prices have risen, as in Changsha County, this has encouraged non-energy intensive production with higher value-added.

Source: World Bank (1997).

Rent taxes. The environment can be a major source of revenue. The potential for additional rent capture is substantial in the forest sector of may countries, and has been estimated to amount to US$9 billion per year. Not all of this can reasonably be captured, due to illegal logging and poor data availability. However, moving towards better rent capture for forestry would dampen the rapid depletion of tropical forests, and could be particularly important for small, forest-rich countries in terms of their fiscal revenue (World Bank, IMF and UNEP, 2002).

Charging visitors fees for visiting protected areas is another under-utilized form of rent capture. Some US$1-3 billion per year could probably be raised in developing countries if fees were raised to levels of actual willingness to pay among visitors. Some of these areas already charge, but many refrain from charging visitors, especially foreign visitors, fees that approach their appreciation for the environmental services provided by protected areas (World Bank, IMF and UNEP, 2002).

Rent taxes are more common for countries with rich fisheries that are exploited by other countries' fishing fleets – such as Japan, Korea, Taiwan and Spain. While most countries in this position do charge for licenses or have joint venture agreements, they are often not receiving the full amount. The size of fishery rent tax revenues from other countries' fleets is significant for certain countries – in particular for small islands in the Pacific and certain African countries. Between 1993 and 1999, Mauritania received 15% of total budget revenue from European Community fishing agreements, 13% for Sao Tome and 30% for Guinea Bissau (IFREMER, 1999).

Pollution charges. It is also important, where possible, to adjust market prices to include the non-marketed environmental effects. Examples include 'green taxes', effluent/emissions fees, deposit refund schemes, tradable permits and so forth. The poverty relevance of these instruments lie primarily in their ability to signal the full social cost of pollution and environmental damage, thereby providing an incentive to limit damaging activities that generally tend to impact the poor most (World Bank, 2000b).

These ideas have been vigorously put into practice in many countries. For example, Bulgaria derives about 15 percent of its tax revenue from environmental fees, and the figure for Hungary is 12 percent. The average for eleven transition economies in Central and Eastern Europe is 2 percent of GDP – a quite substantial contribution to fiscal revenue. China earned US$600 million in 1999 from emission charges. Most of these funds went to finance pollution abatement measures (World Bank, IMF and UNEP, 2002).

Using market-based instruments (MBIs) to ensure that environmental costs are incorporated in market prices is institutionally demanding. A gradual and flexible approach is necessary. Environmental levies are often met with stiff opposition from the polluters who must pay, but earmarking the revenue from environmental fees can improve public acceptance of such levies. A review of the experience of eleven Latin American countries using MBIs emphasize that revenues must be channeled to local authorities so that they can build the institutional capacity required for effective implementation (Huber et al., 1998).

Price reform is important in correcting market signals, but there will always remain some environmental issues that require direct regulation of activities, including outright prohibition, in order to protect the environment and the poor. Examples include the banning of particularly harmful pesticides, and the regulation of allowable applications of others. These measures create an incentive for private producers to find new and more environmentally friendly products that can achieve the same objectives.

2.4 Reforming international and industrialized country policies

Key areas for policy action:
- ☐ Reform trade and industrialized country subsidy policies
- ☐ Make foreign direct investment more pro-poor and pro-environment
- ☐ Enhance the contribution of multilateral environmental agreements to national development objectives
- ☐ Encourage sustainable consumption and production
- ☐ Enhance the effectiveness of development cooperation and debt relief

Reducing poverty through improved environmental management cannot be achieved on a sustainable basis through domestic action alone. Developing countries are economically and socially linked to the rest of the world through trade, investment flows, development co-operation, and humanitarian aid. These capital flows can each have varying impacts on the economic, social, and environmental sustainability of partner countries. There is a growing recognition of the need for more coherent global economic policy frameworks, including the international policies of developed countries, in order to better reflect the concerns of developing countries, sustain the environment and reduce poverty. This implies support for the poverty reduction and sustainable development strategies of developing countries, in particular for domestic policies that enhance sustainable development and create an economic environment conducive to environmentally sustainable trade, investment and economic growth; and it requires international economic frameworks that provide sustainable growth opportunities for developing countries, including market access for developing country exports.

Reform trade and industrialized country subsidy policies

International trade can make a decisive contribution to sustainable development by promoting the equitable integration of developing countries and the poor into the global economy, which can significantly boost economic growth. However, trade and investment liberalization will provide maximum benefit when carried out within a sound supporting domestic policy framework – including pro-poor economic policies – and pursued in tandem with sound environmental management.

In some sectors such as agriculture, many developing countries are still unable to realize their comparative advantage because agricultural trade policies in industrialized countries have the effect of depressing world prices for farm products. Protection in rich countries cost developing countries more than $100 billion per year (World Bank, 2002a). The OECD countries subsidize their agriculture with almost US$1billion per day, much of it encouraging use of agro-chemicals and planting of lands that otherwise would have been left fallow. These subsidies also have the effect of creating barriers to export of agricultural commodities from the poorer countries, making poverty reduction more difficult. Subsidies for marine fisheries have been estimated to total about US$25 billion per year, or about one third of the value of the catch. This contributes significantly to the global pressure on this natural resource (Myers and Kent, 2001).

The exact effect of agricultural trade liberalization by developed countries on the natural resources of developing countries is not clear. More profitable agriculture would encourage both intensification (including wider use of pesticides) and expansion of cropland into forests. Protection of particularly valuable areas would therefore be necessary. However, increased agricultural exports may also lead to more use of fertilizer which can stimulate better ground cover which is important to minimize erosion. It could also lead to crop switching which could be environmentally beneficial. More conclusive impacts arise from international fisheries agreements (for example, by many EC and African states) that often have had adverse development and environmental (resource depletion) impacts on local fisheries communities who depend on fish supply for their food security (MRAG, 2000). These agreements need to be reviewed and reformed.

Furthermore, the trade-related standards of most developed countries can affect developing countries and smaller-scale producers. For instance, legitimate legislation on sanitary and phyto-sanitary (SPS) measures can create challenges for developing countries that often lack the scientific expertise and technical capacity to comply with regulations set by importing developed countries. In effect, SPS measures can create (at least in the short-run) non-tariff barriers that potentially limit the ability of developing countries to access foreign markets for their natural resource exports. However, by increasing the assurance that exports are produced in sustainable ways and that SPS standards are met, such measures can also add value and marketability to products.

This is the case of organic shade-grown coffees that continue to earn fairly high prices despite generally depressed global market prices for lower-grade coffee. The application of certification standards for forest management practices is another promising area. An example of successful adoption of certified sustainable forest management (SFM) and market access is provided by Portico S.A. of Costa Rica. The company manufactures high-end mahogany doors that command a premium price. Thanks to its certified SFM can be exported worldwide without controversy at a time when tropical deforestation is an increasing concern. (Diener, 1998). These environmental standards need to be combined with capacity development in developing countries, in particular among small and medium-sized producers, to allow them to effectively meet the requirements and to turn them into a market advantage rather than an obstacle (see Box 15).

BOX 15: SUCCESSFUL ADJUSTMENT TO ENVIRONMENTAL HEALTH STANDARDS

In 1989, Germany, the leading export market for Indian leather products, banned the import of consumer goods containing PCP and a large number of dyes, citing concerns of human health. These chemicals were routinely used in leather tanning in India. It came as a shock to this important export industry, which ranked fourth in revenue at the time.

The export ban prompted a quick regulatory action by the Indian government to prohibit manufacturing of the banned chemicals, the application of standardized methods for testing, so as to ensure compliance, and rapid development of low-cost substitutes. Surprisingly, this example shows that even highly dispersed, traditional small-firm clusters can successfully meet strict environmental standards in a relatively short time and stay competitive.

Source: Pillai (2000).

Make foreign direct investment more pro-poor and pro-environment

Foreign direct investment (FDI) and foreign portfolio flows now dwarf official development assistance, with over US$160 billion by the end of the last decade

(IMF, World Bank, UNEP, 2002). Even though these flows are focused on a handful of countries, foreign investment is still a key part of resource inflows in the remaining developing countries. Indeed, in order to promote poverty reduction, many countries are seeking to encourage foreign investment. This investment is particularly important to the poverty-environment agenda in those countries where foreign investment is concentrated in resource extraction, infrastructure and manufacturing sectors.

The overall environmental impact of multinational enterprises in developing countries is mixed – while there is not evidence of a "race to the bottom" in terms of environmental standards (World Bank, 2002), there is mixed evidence that foreign firms are cleaner than domestic ones once firm size is included (Zarsky, 1999). However, multinational firms operating in developing countries are increasingly trying to improve environmental performance supported by a number of important initiatives. In 2000, OECD members agreed upon a revised voluntary Code of Conduct for Multinational Enterprises, which has a significant environmental component (OECD, 2000). The UN has been promoting a Global Compact with the private sector that has nine principles, including on the environment. The Global Reporting Initiative, with the support of UNEP, is a multi-stakeholder international undertaking that is drawing up an international standard for reporting on the economic, social and environmental dimensions of a firm's activities, products and services (GRI, 2000).

Foreign direct investment is particularly linked to poverty-environment issues through the oil, gas and mining sectors. Many of the world's poorest countries – Papua New Guinea, Chad, Mozambique, Nigeria – are the site of major investments, with the minerals often located in isolated regions. However, the contribution of an oil, gas or mining corporation to a country's wealth through tax and royalty revenues is frequently not matched by the influence that company has over revenue management. Companies with long-term investments have an incentive to improve relations with local residents. In some cases, this has led to investments in local schools, clinics and infrastructure. Generally, the companies would prefer to see this as the role of national and local governments. The problem arises where governments do not make these investments, and the private companies are reluctant to apply pressure on the host government for fear that they will lose out – for example, by not being awarded future contracts.

Targeted partnerships between investors, the host country national and regional government, development agencies and the local people affected can begin to address these problems. An example is the Lihir gold mine in Papua New Guinea, where participation by local residents as shareholders was financed by an investment bank. Furthermore, a closer alignment of social investment practices among oil companies, municipal governments and development agencies can provide the political incentive to redirect revenues back to the regions where minerals are extracted. Greater complementarity between community development activities of corporations and the regional development plans of municipal authorities can improve the responsiveness of government to community needs and increase the perceived legitimacy of public office (Warner, 2000).

BOX 16: MINING COMPANIES AND THE ENVIRONMENT IN LATIN AMERICA

Detailed studies of the mining sector of Chile, Peru, Brazil and Bolivia during a period of privatization found that environmental damage was not evenly distributed within the minerals sector of each country. Rather, it seemed to vary according to factors such as type of mineral; vintage of technology; stage of investment; stage of operation; level of integration; effectiveness of environmental regulation and its enforcement; and socioeconomic context (including poverty in local communities and work-force education and training). Most of all, environmental performance varied according to the firm's inherent technological dynamism – which did explain foreign firms generally better performance than state-owned national firms.

In the Chilean industry, several international mining firms adopted environmental practices in advance of legislated norms and institutional recommendations. While the state-owned companies face massive challenges in dealing with their sins of the past in terms of accumulated environmental problems, combined with other factors such as the state companies' history, culture, and resource constraints.

However, in Brazil - while foreign firms did sometimes have environmentally proficient practices due to their greater technological capacity and financial resources – others have lagged in the implementation of practices already adopted in the companies' more stringently regulated home countries.

Source: Warhurst (1998).

Enhance the contribution of multilateral environmental agreements to national development objectives

Globalization and global environmental change have focused international attention on the role of global public goods such as biodiversity, the atmosphere, international waters and global agricultural and health research in achieving sustainable development

Two of the major environmental global public goods – a stable climate and maintenance of biodiversity – have many benefits for the poor. The main historic responsibility for climate change lies with the developed world. However, the developing world includes countries whose increasing emissions of greenhouse gases and related pollutants are unhealthy, unsustainable and less and less likely to be the least cost options for development. A binding international agreement must be found that effectively and equitably reduces the emission of greenhouse gases. Another area is to ensure that multilateral assistance is used to help minimize (and does not increase) greenhouse gas emissions and related pollutants, and to use such assistance to implement win-win solutions that advance several development goals at once, such as public health, biodiversity conservation and climate change mitigation.

The causes of biodiversity loss are more complex than climate change. As the whole world benefits from maintaining biodiversity, and developing countries lack resources, it is incumbent on the developed world to bear a fair proportion of the costs of global biodiversity conservation, both through direct assistance and through more careful assessment of the impact of their trade, investment and other interactions with the developing world. A major instrument for direct assistance is the Global Environment Facility (GEF). Negotiations are currently ongoing for the GEF's next financing period, with a significant increase required to help protect the world's climate and biodiversity and other global environmental goods that benefit all, but often the poor most of all.

Over the past 50 years, international environmental policies have been agreed in the context of numerous multilateral environmental agreements (MEAs). Each agreement has customarily been designed to address a pressing environmental issue. As a result, some agreements support and strengthen the aims of others, while others have objectives that overlap and contradict one another. There is a need for better coordination and harmonization between MEAs during negotiation and implementation stages, and for elimination of contradictions and overlaps. Increased synergy among MEAs is required for more efficiency and ensure that MEAs contribute to furthering their own objectives while ensuring that the environmental considerations of such agreements are integrated in the broader dimensions of sustainable development and do not contradict other legal regimes.

Developing countries should be enabled to take on increased responsibilities under global agreements to which they are party, and to ensure that these agreements adequately reflect their concerns. Effective participation in international negotiations, however, requires capacity and resources that are often lacking in the poorest countries. It also requires political will for the interests of the poor to be made central to both the negotiation and implementation of these MEAs. Developed countries should assist developing countries in implementing the objectives of the MEAs to which they are each parties, and take care to ensure that they do not unilaterally, or through multilateral operations, support actions of developing countries that are not in compliance with MEAs to which they are party.

Encourage sustainable consumption and production

Developed country consumers and producers through their trade, investment, pollution emissions and other activities affect the environmental conditions of developing countries. While this chapter focuses primarily on specific steps relevant to trade, investment and global public goods, there is a broader underlying issue – the level of production and consumption in the industrialized world.

Making rich country consumption and production more sustainable will require a complex mix of institutional changes – addressing market and government failures as well as broad public attitudes. As in developing countries, it will also require working with many different stakeholders in government, civil society and the private sector. Again as in developing countries, it is not just a technical process – but a political one – with certain groups that will welcome change, while others will resist it. One interesting example of the new alliances that are being forged both between stakeholders in developed countries and their partners in developing countries is the recent Memorandum of Understanding between Indonesia and the UK on Indonesia forestry exports (see Box 17).

BOX 17: CURBING DEVELOPED COUNTRY IMPORTS OF ILLEGAL TIMBER FROM INDONESIA

Indonesia is a major exporter of timber to Europe. Much of this timber is illegally and/or unsustainably harvested. In 2001, a conference in Asia on illegal logging examined how both developing country producers and developed country consumers could work together to promote sustainable logging. In 2002, this led to a Memorandum of Understanding between the Indonesian Minister of Forestry and UK Ministers for the Environment and International Development to cooperate on forest law enforcement and combat illegal logging and trade.

This agreement will help set up legal compliance for Indonesian forest exports, which will eventually allow all UK imports to be only from legal sources. This would require amending UK customs law, which may also require EU legislation. In the meantime, the UK Timber Trade Federation has already drawn up a voluntary code of practice to work with Indonesian suppliers towards the elimination of illegal logging. G-8 partners, including the US, Germany and Japan are interested, and the EC is convening a meeting to debate extending the arrangement to the whole of the European Union. An African conference on illegal logging is now being planned between African producers and the US, France, UK and the EC.

Source: Internal DFID documents.

The rich countries of the world recently acknowledged their responsibility to reduce environmental pressure in the OECD report Sustainable Development, Critical Issues (OECD, 2001): "OECD countries have a key role to play in addressing the pressures on the environment from human activities. With 18% of the world's population, they account for over half of today's total energy consumption, over 60% of cereals consumption, 31% of consumption of food fish, 44% of consumption of forest products and a large fraction of the cumulative damage imposed on the environment globally." The OECD report goes on to identify detailed steps in the energy, transport, agriculture and manufacturing sectors to reduce environmental damage – which will benefit both OECD countries, but also developing countries. For each of these key sectors, the OECD report provides a detailed list of institutional, regulatory and economic policy reforms to reduce environmental damage in its 30 member states. The OECD also carries out regular 'peer reviews' of its member states to assess environmental performance. These are Ministerial-level reviews and the final reports are public documents which provide constructive suggestions for improvement.

The EC also has been explicit in its strategy for the 15 members of the European Union: "Industrialized countries have important responsibilities in promoting sustainability initiatives – first and foremost by putting their own house in order, and by supporting a move to sustainable production and consumption patterns; in addition by ensuring more consistent market opening, increased public and private financing of development cooperation, as well as better functioning and greater stability in the international financial system" (EC, 2002).

Enhance the effectiveness of development cooperation and debt relief

To achieve the Millennium Development Goal of halving absolute poverty by 2015 will require an approximate doubling of official development assistance (Devarajan et al., 2002; Zedillo Report, 2001). This would still only bring the total level of aid to less than half a percent of GNP in OECD countries, still far below the internationally acclaimed goal of reaching 0.7 percent of GNP. To eradicate poverty will demand a much more ambitious effort, and the financial flows must be received with efficiency and accountability to be effective. International aid works in a good policy environment.

Many developing countries are burdened by unsustainable levels of debt. This hampers economic growth and undermines the ability of some countries to provide health, education and other basic services for their population. When unsustainable debt leads to budgetary cuts, environmental administration and services often are a target, leading to a slackening of environmental management. The Heavily Indebted Poor Country (HIPC) Initiative aims to tackle the problem of unsustainable debt, and to ensure that the benefits from debt relief are used to reduce poverty and to avoid entering into a renewed spiral of indebtedness. Debt-for-nature swaps are another potential means for addressing poverty reduction and environmental management objectives.

Aid and debt relief can be provided to help governments make many of the policy changes recommended in this paper. As in developing countries, development cooperation agencies are seeking to improve their governance structures and operational effectiveness by:

☐ Adopting a more explicit commitment to poverty reduction as the over-riding objective of development cooperation;

☐ Putting developing countries in the drivers' seat through support of nationally-owned processes and improved aid coordination;

☐ Ensuring greater transparency, and greater engagement with civil society, at both policy and operational levels;

☐ Making development cooperation more results-based and accountable by focusing more strongly on development outcomes, in particular by strengthening capacity to help countries achieve the Millennium Development Goals;

☐ Decentralizing operations and empowering country-level staff to be more flexible and responsive to country needs.

To help move the poverty-environment agenda forward, development agencies must learn from past mistakes and incorporate these lessons into the new context for development cooperation. The shift in development cooperation to focus more explicitly on poverty reduction and greater country ownership provides new opportunities for improving environmental management. While our agencies have committed themselves to better environmental management as a tool for poverty reduction, this now has to be operationalized throughout our respective organizations – both in headquarters and in country offices. We also need to place less emphasis on our own in-house procedures for evaluating environmental risks – although these will remain important – and much more emphasis on helping developing the capacity of our partners to build up their own national processes to take up opportunities to reduce poverty through better environmental management.

Putting these commitments into practice requires major changes in the way we do business – but we cannot afford to fail. To take this message forward will require improved staff training and staff skills. New tools and procedures need to be implemented. The shift in aid towards more upstream work with financial support provided for whole sectors and budgets provides new challenges. The traditional environmental impact assessment approach needs to be moved up to sectors and policies, but also made more focused on environmental issues that affect the poor – and integrated with poverty assessments. There is a need to provide incentives to programme managers to mainstream poverty-environment issues. Senior management needs to provide strong leadership – not just in policy statements, but also in the way resources and staff are allocated. Finally, there is a need for effective and transparent monitoring of progress against stated objectives and targets.

Conclusion

This paper set out to articulate ways to reduce poverty in a sustainable manner through better environmental management. We have mapped out the key relationships between environment and poverty. Specifically, we have pointed to the enormous burden of disease that impacts the poorest through polluted water and air. We have also illustrated how directly and heavily dependent the poor are on natural resources and ecosystem services, and how their degradation can undermine their livelihoods. Related to this point is the vulnerability to environmental disasters that the poor are exposed to, and their limited ability to cope with such shocks. We know this not only because of empirical evidence, but most compellingly through what the poor say themselves.

While many links between environment and poverty are reasonably clear, we have also held up relationships that are controversial. Environment and growth, environment and population, natural resource degradation and the poor, are all themes that have been subject to much generalization and over-simplification. Effective solutions must be guided by a nuanced understanding of the specifics of these relationships, often determined by localized institutions and policies.

While we share a sense of urgency in combating environmental degradation, we have not dwelled at length on descriptions of problems that are generally, albeit not universally, agreed. Instead, we have emphasized links between poverty and environment, and above all, what lessons we can learn for the future. Hence, this paper is one that looks ahead with some degree of hope and optimism for the future: there are sometimes win-win opportunities, and there are rational ways of dealing with tradeoffs. Environmental degradation is not inevitable, nor is it an unavoidable sacrifice on the altar of economic growth. On the contrary, better environmental management is key to poverty reduction.

In that spirit, this paper has discussed a large set of measures, both at the national and international level, which can be taken to reduce poverty and enhance environmental quality. This has taken us outside the realm of narrowly conceived "environmental management", as the links between poverty and environment are complex and crosscutting. We have not attempted to be comprehensive and provide detailed recommendations. The details are best left to inclusive national processes for shaping poverty reduction and sustainable development strategies. Rather, we have tried to be selective and strategic; focusing on the key items around which we hope to stimulate debate and action.

The WSSD is an opportunity for us all to focus on what is most important and to forge agreements that can lead the way forward. There can be no more important goal than to reduce and ultimately exterminate poverty on our planet.

Endnotes

Abbreviations and Acronyms

CBD   Convention on Biological Diversity
CGIAR   Consultative Group on International Agricultural Research

DAC   Development Assistance Committee (OECD)
DALYs   Disability-Adjusted Life Years
DFID   Department for International Development (UK)
EC   European Commission
EIA   Environmental Impact Assessment
FAO   Food and Agriculture Organization
FDI   Foreign Direct Investment
GDP   Gross Domestic Product
GEF   Global Environment Facility
HIPC   Heavily Indebted Poor Country Initiative
IFAD   International Fund for Agricultural Development
IIED   International Institute for Environment and Development
IPCC   International Panel on Climate Change
IPRSP   Interim Poverty Reduction Strategy Paper
IUCN   World Conservation Union
MEA   Multilateral Environmental Agreement
NGO   Non-Governmental Organization
NSSD   National Strategy for Sustainable Development
OECD   Organization for Economic Co-operation and Development
PEI   Poverty and Environment Initiative (UNDP)
PPA   Participatory Poverty Assessment
PRS   Poverty Reduction Strategy
PRSP   Poverty Reduction Strategy Paper
SEA   Strategic Environmental Assessment
SIA   Social Impact Analysis
UNCCD   United Nations Convention to Combat Desertification and Drought
UNCED   United Nations Conference on Environment and Development
UNCHS   United Nations Centre for Human Settlements (Habitat)
UNDP   United Nations Development Programme
UNDESA   United Nations Department for Economic and Social Affairs
UNEP   United Nations Environment Programme
UNFCC   United Nations Framework Convention on Climate Change
WBCSD   World Business Council for Sustainable Development
WHO   World Health Organization
WRI   World Resources Institute
WSSD   World Summit on Sustainable Development
WWF   World Wide Fund for Nature

References

ACTS (African Centre for Technology Studies). 2000. Ecological Sources of Conflict in Southern Africa. Discussion Note. Nairobi.

Agarwal, A. and S. Narain. 1999. Community and Household Water Management: The Key to Environmental Regeneration and Poverty Alleviation. Poverty and Environment Initiative Background Paper 2. UNDP, New York.

Arnold, J. and P. Bird. 1999. Forestry and Poverty. Poverty and Environment Initiative Background Paper 6. UNDP, New York.

Ashley, C., D. Roe and H. Goodwin. 2001. Pro-poor Tourism Strategies: Making Tourism Work for the Poor: A Review of Experience. Pro-poor tourism report No. 1. Overseas Development Institute, International Institute for Environment and Development and Centre for Responsible Tourism.

ADB (Asian Development Bank). 2000. Privatization of Water Supplies in Ten Asian Cities. Manila.

Bann, C. 1997. An Economic Analysis of Alternative Mangrove Management Strategies.

Bartley, D.M. 1999. Species Introduction, International Conventions, and Biodiversity: Impacts, Prospects and Challenges. In Subasinghe, Arthur, Philips and Reantaso (eds.), Thematic Review on Management Strategies for Major Diseases in Aquaculture. Report of a workshop in Cebu, Philipppines.

Bass, S., M. Thornber, S. Roberts and M. Grieg-gran. 2001. Certification's Impacts on Forests, Stakeholders and Supply Chains. IIED (International Institute for Environment and Development), London.

Bojö, J., J. Bucknall, K. Hamilton, N. Kishor, C. Kraus and P. Pillai. 2001. Environment. In Poverty Reduction Strategy Sourcebook. World Bank, Washington, D.C.

Bojö, J. and S. Pagiola. 2000. Natural Resources Management. A contribution to the World Bank Environment Strategy. World Bank, Washington, D.C.

Bojö, J. and R.C. Reddy. 2002. Poverty Reduction Strategies: A Review of 40 Interim and Full PRSPs. Environment Department Paper No. (x). World Bank, Washington, D.C.

Boyd, C. and T. Slaymaker. 2000. Re-examining the "more people, less erosion" hypothesis: special case or wider trend. ODI Natural Resource Perspectives No. 63. Overseas Development Institute, London.

Brocklesby, M.A. and E. Hinshelwood. 2001. Poverty and the Environment: What the Poor Say. Centre for Development Studies, University of Wales.

Bruce N., L. Neufeld, E. Boy and C. West. 1998. Indoor biofuel air pollution and respiratory health: the role of confounding factors among women in highland Guatemala. International Journal of Epidemiology, 27: 454-58.

Brunner, J., F. Seymour, N. Badenoch and B. Ratner 2000. Forest Problems and Law Enforcement in Southeast Asia: The Role of Local Communities. WRI, Washington, D.C.

Campbell, H. 1997. Indoor air pollution and acute lower respiratory infections in young Gambian children. Health Bulletin, 55: 20-31.

Cavensdish, W. 1999. Empirical Regularities in the Poverty-Environment Relationship of African Rural Households. Center for the Study of African Economies. Working Paper Series 99-21. London.

CGIAR (Consultative Group on International Agricultural Research). 1997. Report of the Study on CGIAR Research Priorities for Marginal Lands. Technical Advisory Committee Working Document, TAC Secretariat. FAO, Rome.

Commission on Sustainable Development. 2002. Implementing Agenda 21: Report of the Secretary-General. Advance Unedited Text (E/CN.17/2002/PC.2/....). United Nations, New York.

Dasgupta and Das. 1998. Health Effects of Women's Excessive Work Burden in Deforested Rural Areas of Uttarkhand. Paper presented at the National Conference on Health and Environment. Center for Science and Environment, New Delhi.

Devarajan, S., M.J. Miller and E. Swanson. 2002. Goals for Development: History, Prospects, and Costs. World Bank, Washington, D.C.

DFID (Department for International Development). 2000a. Achieving sustainability: poverty elimination and the environment. Strategies for achieving the international development targets. London.

___. 2000b. Integrating sustainability into PRSPs: the case of Uganda. Environmental Policy Department. London.

___. 2000c. Strategies for sustainable development: can country-level strategic planning frameworks converge to achieve sustainability and eliminate poverty? DFID Background Briefing. London.

___. 2001a. Biodiversity – a crucial issue for the world's poorest. London.

___. 2001b. Making government work for poor people. Strategies for achieving the international development targets. London.

___. 2002a. Draft Land Policy Paper. London.

___. 2002b. Review of the Inclusion of Poverty-Environment Issues in Selected Poverty Reduction Strategy Papers and Joint Staff Assessments. Environmental Policy Department. London.

Diener, B.J. 1998. Portico, S.A. Strategic Decisions 1982-1997. In, The Business of Sustainable Forestry: Case Studies. A Project of the Sustainable Forestry Working Group.

Duraiappah, A.K., G. Ikiara, M. Manundu, W. Nyangena, R. Sinanga, H. Amman and D. Lamba. 2000. CREED Policy Brief. Poverty-Environmental Degradation Nexus, Mazingira Institute, Nairobi, Kenya.

EC (European Commission). 2001. Commission Staff Working document SEC(2001) 609 of 10 April 2001 on Integrating the environment into EC economic and development co-operation, and the respective Council Conclusions of 31 May 2001.

___. 2002. Towards a Global Partnership for Sustainable Development, Communication from the Commission to the European Parliament, the Council, the Economic and Social Committee and the Committee of the Regions, 13 February 2002, COM(2002) 82 (final), Brussels.

Ekbom, A. and J. Bojö. 1999. Poverty and Environment: Evidence of Links and Integration into the Country Assistance Strategy Process. Environment Group, Africa Region, World Bank. Washington, D.C.

ERM (Environmental Resources Management). 1997. Evaluation of the Environmental Performance of EC Programmes in Developing Countries. Brussels.

ESD (Energy for Sustainable Development). 2000 Poverty reduction aspects of successful improved household stoves programme DFID knowledge and research paper. London.

Ezzati, M, and D.M. Kammen. 2001. Indoor air pollution from biomass combustion and acute respiratory infections in Kenya: an exposure-response study. The Lancet, 358(9282).

Gilbert, R. and A. Kreimer. 1999. Learning from the World Bank's Experience of Natural Disaster Related Assistance. Disaster Management Facility, Urban and Local Government, World Bank, Washington, D.C.

Giovannuci, D. 2001. Sustainable Coffee Survey of the North American Specialty Coffee Industry. Conducted for The Summit Foundation, The Nature Conservancy, North American Commission for Environment Cooperation, Specialty Coffee Association of America, and the World Bank (available at http://www.scaa.org).

GEF (Global Environment Facility). 1998. GEF Evaluation of Experience with Conservation Trust Funds. GEF/C.12/Inf.6. Washington, D.C.

Global Witness. 2001 and 2000. A number of references are provided on the Global Witness web site (http://www.oneworld.org/globalwitness/).

Goldman, L. and N. Tran. 2002. Toxics and Poverty. World Bank, Washington, D.C. Processed.

GRI (Global Reporting Initiative). 2000. Sustainability Reporting Guidelines on Economic, Environmental, and Social Performance. Boston, Massachusetts.

GRID/Arendal. 1997. Mapping Indicators of Poverty in West Africa. DEIA/TR.97-8. Technical Advisory Committee Working Document, CGIAR, FAO, Rome.

Hardoy J., D. Mitlin and D. Satterthwaite. 2001. Environmental Problems in an Urbanising World. Earthscan Publications, London.

Hazell, P. and J.L. Garrett. 1996. Reducing Poverty and Protecting the Environment: The Overlooked Potential of Less-favored Lands. 2020 Brief 39. IFPRI (International Food Policy Research Institute), Washington, D.C.

Heinrich Böll Foundation. 2002. The Jo'burg Memo: Fairness in a Fragile World. Memorandum for the World Summit on Sustainable Development. Washington, DC. [check reference is correct]

Henninger, N. and A. Hammond. 2000. A Strategy for the World Bank: Environmental Indicators Relevant to Poverty Reduction. World Resources Institute, Washington, D.C.

Hodess, R. (ed.). 2001. Global Corruption Report 2001. Transparency International, Berlin.

Huber, R.M., J. Ruitenbeek and R. Seroa da Motta. 1998. Market Based Instruments for Environmental Policymaking in Latin America and the Caribbean: Lessons from Eleven Countries. World Bank Discussion Paper No. 381. Washington, D.C.

Iannariello, M.P., P. Stedman-Edwards, R. Blair and D. Reed. 2001. Environmental Impact Assessment for Macroeconomic Reform Programs. Macroeconomics Program Office, World Wildlife Fund, Washington, D.C.

ICRC (International Committee of the Red Cross). 1999. Annual Report. Geneva.

___. 2001. World Disasters Report 2000. Focus on Recovery. International Federation of Red Cross and Red Crescent Societies, Geneva.

IFAD (International Fund for Agricultural Development). 2001. Rural Poverty Report 2001. Rome.

IFREMER. 1999. Evaluation of Fisheries Agreements Concluded by the European Community. Summary Report.

IIASA (International Institute for Applied Systems Analysis). 2001. Global Agro-ecological Assessment for Agriculture in the 21st Century. Vienna.

IIED (International Institute for Environment and Development). 2000. Evaluating Eden: Exploring the myths and realities of community-based wildlife management. Evaluating Eden Series No. 8. London.

___. 2000. Rural Livelihoods and Carbon Management. IIED Natural Resource Issues Paper. London.

___. 2002. Drawers of Water II. In collaboration with Community Management and Training Services Ltd. (Kenya), Institute of Resource Assessment of the University of Dar es Salaam (Tanzania), Department of Pediatrics and Child Health of Makerere University Medical School (Uganda). London.

IIED and WBCSD (World Business Council for Sustainable Development). 2002. Mining, Minerals and Sustainable Development (Draft Report). London.

IMF (International Monetary Fund), World Bank and UNEP (United Nations Environment Programme). 2002. Financing for Sustainable Development: An input to the World Summit on Sustainable Development. Revised Consultation Draft.

IPCC (International Panel on Climate Change). 2001. Summary for Policymakers. Climate Change 2001:Impacts, Adaptation, and Vulnerability. A Report of Working Group II of the IPCC. Geneva.

ISDR Secretariat (International Strategy for Disaster Reduction). 2002. Disaster Risk and Sustainable Development: understanding the links between development, environment and natural hazards leading to disasters. Draft Background Document for the World Summit on Sustainable Development (revised version 14 April 2002). Bonn.

Jodha, N.S. 1986. Common property resources and rural poor in dry regions of India. Economic and Political Weekly, 21(27): 1169-1181.

Johnstone, N., L. Wood and R. Hearne. 1999. The Regulation of Private Sector Participation in Urban Water Supply and Sanitation: Realising Social and Environmental Objectives in Developing Countries. Discussion Paper 99-01, Environmental Economics Programme, International Institute for Environment and Development, London.

Koziell, I. and J. Saunders. 2001. Living off Biodiversity: Exploring Livelihoods and Biodiversity. Issues in Natural Resources Management. IIED, London

Leighton, M. 1999. Environmental Degradation and Migration. In Drylands, Poverty and Development. Proceedings of the World Bank Round Table. World Bank, Washington, D.C.

Lele, S. 2001. Godsend, slight of hand or just muddling through: Joint water and forest management in India. ODI Natural Resource Perspectives. Overseas Development Institute, London.

Lemos, M. 1998. The politics of pollution control in Brazil: state actors and social movements cleaning up Cubatao. World Development, 26(1): 75-87.

Loftus, A.J. and D. McDonald. 2001. Of liquid dreams: a political ecology of water privatization in Buenos Aires. Environment & Urbanization, 13(2): .

Lvovsky, K. 2001. Health and Environment. Environment Strategy Papers, No. 1. Environment Department, World Bank, Washington, D.C.

MRAG (Marine Resources Advisory Group). 2000. The impact of fisheries subsidies on developing countries. Prepared in association with Cambridge Resource Economics and International Institute for Environment and Development. London.

Morris, E. and S.C. Rajan. 1999. Energy as it Relates to Poverty Alleviation and Environmental Protection. Poverty and Environment Initiative Background Paper 4. UNDP (United Nations Development Programme), New York.

Munasinghe, M. and W. Cruz. 1995. Economy-wide Policies and the Environment: Lessons from Experience. World Bank, Washington, D.C.

Munasinghe, J. Warford, A. Schwab, W. Cruz and S. Hansen. 1994. The Evolution of Environmental Concerns in Adjustment Lending: A Review. World Bank, Washington, D.C.

Murray, C. and A. Lopez. 1996. The Global Burden of Disease. Harvard University Press, Cambridge, MA.

Myers, N. and J. Kent. 2001. Perverse Subsidies: How Tax Dollars Can Undercut the Environment and the Economy. Island Press, Washington, D.C.

Narayan, D., with R. Patel, K. Schafft, A. Rademacher and S. Koch-Schulte. 2000. Voices of the Poor: Can Anyone Hear Us? World Bank, Washington, D.C.

Nickson, A. and R. Franceys. 2001. Tapping the Market: Can Private Enterprise Supply Water to the Poor? ID21.

Nordhaus and Kokklenberg (eds.). 2001. Nature's Numbers: Expanding the National Economic Accounts to Include the Environment. National Research Council, National Academy Press, Washington, D.C.

Nunnan, F. et al. 2001. Poverty-Environment Linkages: Developing and Assessing the Use of Indicators. A report to DFID (Department for International Development). London.

OECD (Organization for Economic Cooperation and Development). 2000. The OECD Guidelines for Multinational Enterprises, Revision 2000. Paris.

___. 2001. Poverty-Environment Linkages. Working Party on Development Cooperation and the Environment (February 14, 2001). Paris.

___. 2001. Sustainable Development, Critical Issues. Paris.

___. 2001. The DAC Guidelines. Strategies for Sustainable Development: Guidance for Development Cooperation. Paris.

Ostrom, E. 1990. Governing the Commons: the Evolution of Institutions for Collective Actions. Cambridge University Press, Cambridge.

Osuntogun, A. 2002. Applied Poverty-Environment Indicators: The Case of Nigeria. Report submitted to the Environment Department, World Bank. Abuja. Processed.

Oxfam. 2002. Poverty in the midst of Wealth: The Democratic Republic of Congo. Oxfam Policy Papers, Oxfam Briefing Paper. Oxford.

Pillai, P. 2000. The State and Collective Action: Successful Adjustment by the Tamil Nadu Leather Cluster to German Environmental Standards. Unpublished MCP Thesis, Massachusetts Institute of Technology.

Read, T. and L. Cortesi. 2001. Stories at the Forest Edge: The KEMALA Approach to Crafting Good Governance and Sustainable Futures. Biodiversity Support Program, Washington, D.C.

Reed, D. (ed.). 1992. Structural Adjustment and the Environment. Earthscan Publications, London.

Reed, D. (ed.). 1996. Structural Adjustment, the Environment and Sustainable Development. Earthscan Publications, London.

Reed, D. and H. Rosa. 1999. Economic Reforms, Globalization, Poverty and the Environment. Poverty and Environment Initiative Background Paper 5. UNDP (United Nations Development Programme), New York.

Reij, C. and A. Waters-Bayer (eds.). 2001. Farmer Innovation in Africa. Earthscan Publications, London.

Satterthwaite, D. 1999. Links Between Poverty and the Environment in Urban Areas of Africa, Asia and Latin America. Poverty and Environment Initiative Background Paper 1. UNDP (United Nations Development Programme), New York.

Scherr, S. 1999. Poverty-Environment Interactions in Agriculture: Key Factors and Policy Implications. Poverty and Environment Initiative Background Paper 3. UNDP (United Nations Development Programme), New York.

Shively, G. 2001. Poverty, consumption risk and soil conservation. Journal of Development Economics, 65: 267-290.

Shyamsundar, P. 2002. Poverty-Environment Indicators. Environment Department Paper No. 84. World Bank, Washington, D.C.

Smith, K.R. 1999. Pollution Management in Focus. Indoor Air Pollution Discussion Paper No. 4, Environment Department, World Bank. Washington, DC.

Smith, K.R. 2000. National Burden of Disease in India from Indoor Air Pollution. In Proceedings of the National Academy of Sciences of the United States of America, PNAS 2000 97: 13286-13293.

Songsore, J. and G. McGranahan. 1993. Environment, Wealth and Health: Towards an analysis of intra-urban differentials within the Greater Accra Metropolitan Area, Ghana. Environment and Urbanization, 5(2): 10–34.

Third World Network. 2001. International Environmental Governance: some issues from a developing country perspective. Working Paper prepared for the Chairman of G77. Penang, Malaysia.

Tiffen, M., M. Mortimore and F. Gichuki. 1994. More People, Less Erosion: Environmental Recovery in Kenya. Chichester, New York.

UNCHS (United Nations Centre for Human Settlements). 1996. An Urbanizing World: Global Report on Human Settlements. Oxford University Press, Oxford.

UNDESA (United Nations Department for Economic and Social Affairs). 2002. Guidance in Preparing a National Sustainable Development Strategy: Managing

Sustainable Development in a New Millennium. WSSD Second Preparatory Session, Background Paper No. 13 (DESA/DSD/PC2/BP13). New York.

UNDP (United Nations Development Programme). 1995. Human Development Report 1995. Oxford University Press, New York and Oxford.

___. 1997. Reconceptualising Governance. New York.

___. 1999a. A Better Life…With Nature's Help: Success Stories. Poverty and Environment Initiative. New York.

___. 1999b. Attacking Poverty While Protecting the Environment: Towards Win-win Policy Options. J. Ambler (ed.). Poverty and Environment Initiative Synthesis Paper. New York.

___. 2001. Disaster Profiles of the Least Developed Countries. Emergency Response Division, UNDP, New York.

UNDP, UNDESA (United Nations Department for Economic and Social Affairs) and World Energy Council. 2000. World Energy Assessment. UNDP, New York.

UNEP (United Nations Environment Programme). 2002. Global Environment Outlook 3: Past, Present and Future Perspectives. Earthscan Publications, London.

Waddell, S. 2001. Emerging Models for Developing Water Systems for the Rural Poor: From Contracts to Co-production. Business Partners for Development Water and Sanitation Cluster, London.

Warhurst, A. (ed.). 1998. Mining and the Environment: Case Studies from the Americas. International Development Research Center, Ottawa.

Warner, M. 2000. Tri-sector partnerships for social investment within the oil, gas and mining sector – an analytical framework. Working Paper No 2, Business Partners for Development.

WHO (World Health Organization). 1997. World Health Report. Geneva.

World Bank. 1992. World Development Report 1992: Development and the Environment. Oxford University Press, Oxford.

___. 1996. Energy for Rural Development in China. ESMAP (Energy Sector Management Assistance Programme), World Bank, Washington, D.C.

___. 1996. Pakistan: Economic Policies, Institutions, and the Environment. Report No. 15781–PAK, Agriculture and Natural Resources Division, Country Department I, South Asia Region, Washington, D.C.

___. 1997. Five years after Rio: Innovations in Environmental Policy. Environmentally Sustainable Development Studies and Monographs Series No. 18. Washington, D.C.

___. 1998. Assessing Aid: What Works, What Doesn't, and Why. A World Bank Policy Research Report. Oxford University Press, Oxford.

___. 2000a. A Review of the World Bank's 1991 Forest Strategy and Its Implementation. Operations Evaluation Department, Washington, D.C.

___. 2000b. Greening Industry: New Roles for Communities, Markets, and Governments. Washington, D.C.

___. 2000c. The Quality of Growth. Oxford University Press, Oxford.

___. 2001a. Economic Causes of Civil Conflict and their Implications for Policy. Washington, D.C.

___. 2001b. Engendering Development: Through Gender Equality in Rights, Resources, and Voice. A World Bank Policy Research Report. Oxford University Press, Oxford.

___. 2001c. Making Sustainable Commitments: An Environment Strategy for the World Bank. Washington, D.C.

___. 2001d. Poverty Reduction Strategy Sourcebook. Washington, D.C.

___. 2001e. World Development Indicators. Washington, D.C.

___. 2001e. World Development Report 2000/2001 Attacking Poverty. Washington, D.C.

___. 2002a. Globalization, Growth, and Poverty: Building an Inclusive World Economy. A World Bank Policy Research Report. Oxford University Press, Oxford.

___. 2002b. Water Resources Strategy: Strategic Directions for World Bank Engagement. Draft for discussion (March 25, 2002). Washington, D.C.

___. 2002c. World Development Indicators 2002. Washington, D.C.

WRI (World Resources Institute). 1996. World Resources 1996-1997. The Urban Environment. Oxford University Press, Oxford.

___. 2000. World Resources 2000-2001. People and Ecosystems. Washington, D.C.

WTO (World Trade Organization). 1998. Report of the Appellate Body, United States – Import Prohibition of Certain Shrimp and Shrimp Products, WT/DS58/Appellate Body/R (October 12, 1998).

___. 2001. Ministerial Declaration on 14 November 2001 of the Ministerial Conference Fourth Session, November 9-14, 2001. Doha.

World Water Council. 2000. A Water Secure World. Vision for Water, Life and the Environment. World Water Vision Commission Report. Marseille.

Yaron, G. and White, J. 2002. Mainstreaming cross cutting themes in programme and sector aid: the case of environmental issues. ODI Natural Resource Perspectives, No. 77. Overseas Development Institute, London.

Zarsky, L. 1999. "Havens, Halos and Spaghetti: Untangling Evidence about Foreign Direct Investment and the Environment." In Foreign Direct Investment and the Environment. Organization for Economic Cooperation and Development, Paris.

Zedillo, E.Z., A. Y. Al-Hamad, D. Bryer, M. Chinery-Hesse, J. Delors, R. Grynspan, A.Y. Livshits, A.M. Osman, R. Rubin, M. Singh, and M. Son. 2001. Technical Report of the High-Level Panel on Financing for Development. United Nations, New York.

www.ingramcontent.com/pod-product-compliance
Lightning Source LLC
Chambersburg PA
CBHW081050170526
45158CB00006B/1921